ONE SHOT, ONE KILL

A HISTORY OF THE SNIPER

Andy Dougan

WILLIAM
COLLINS

William Collins
An imprint of HarperCollins*Publishers*
1 London Bridge Street
London SE1 9GF
www.WilliamCollinsBooks.com

First published in Great Britain as *The Hunting of Man*
by 4th Estate in 2004
This William Collins paperback edition published in 2016

1

A catalogue record for this book is
available from the British Library

ISBN 978-0-00-818940-2

Printed and bound in Great Britain by
Clays Ltd, St Ives plc

MIX
Paper from
responsible sources
FSC™ C007454

FSC™ is a non-profit international organisation established to promote the
responsible management of the world's forests. Products carrying the FSC
label are independently certified to assure consumers that they come from
forests that are managed to meet the social, economic and ecological needs
of present or future generations, and other controlled sources.

Find out more about HarperCollins and the environment at
www.harpercollins.co.uk/green

For Findlay

CONTENTS

'There is no hunting like the hunting of man, and those who have hunted armed men long enough, and liked it, never care for anything else thereafter.'

Ernest Hemingway, 'On the Blue Water', *Esquire*, April 1936

PROLOGUE:
THE FIRST SHOT

The city of Lichfield in Staffordshire nestles in rural tranquillity in the very heart of England, just under twenty miles north-east of the sprawling conurbation that Birmingham has become. It has earned the description of 'the loyal and ancient city': ancient because there has been a cathedral church in the city since the year 700, loyal because it remained faithful to the Crown during the Great Rebellion of the mid-seventeenth century.

Modern Lichfield is a hybrid. It has its supermarkets and its chain stores, but unlike in so many town centres these are incorporated into streets that have barely changed in hundreds of years. Timber-framed buildings jostle cheek by jowl for space with modern red brick and steel. Some street names may have changed, but it is still laid out roughly in the same proportions in which it was first fortified by Bishop Roger de Clinton, who looked after the spiritual health of the city from 1129 to 1148.

The landscape of Lichfield is dominated by its famous cathedral. It is the only cathedral in England with three spires, and these imposing landmarks can

1

be seen for miles. There had been one or possibly two Saxon churches on the site since the year 700, but nothing of these remains. The present cathedral is Norman, built at the end of the eleventh century from local sandstone and then extended and finally completed in 1330. Unlike many English cathedrals, Lichfield's had no monastic community attached to it and so possessed no accompanying monastery or cloisters. But a significant population of bishops, priests and other clergy required housing, and their accommodation became known as the Cathedral Close. Bishop de Clinton fortified it in the twelfth century with a great stone wall around the perimeter and a deep ditch dug in front of that as an additional deterrent. Just over two hundred years later, one of his successors, Walter de Langton, completed the cathedral with the construction of the Lady Chapel and also had built for himself a splendid Bishop's Palace. This was replaced in 1687 by the building that currently occupies the site.

A modern visitor to Lichfield would not find it too difficult to cast the mind back seven hundred years. This old part of the city is still laid out in the pattern originally devised by Bishop de Clinton. Dam Street, a collection of buildings of varying vintage, some of them leaning at alarming angles, houses modern businesses, but it still meanders its narrow way down towards the cathedral. Lean against a door lintel in Dam Street now and look towards the great spires; just a little imagination conjures one of the decisive scenes in English history, complete with the acrid whiff of black powder, the yells of the mob, and the screams of the wounded as Lichfield

2

found itself at the centre of the English Civil War.

In 1642 the simmering row between Parliament and the Crown over who should govern the country erupted into the First Civil War. King Charles I raised his standard at Nottingham on 22 August 1642 in what is generally agreed to be the official beginning of hostilities. The banner was blown flat very soon afterwards by a strong wind, taken by many to be an ill omen for the monarch and his Royalist forces, the Cavaliers.

The Civil War divided the country. There were a great many different reasons for choosing sides. Some fought out of loyalty to the Crown, others for conscience, still others because of family connections. Whatever the motive for taking up arms the conflict was generally drawn up on geographical lines. Those for the king included the North of England, Wales, the Borders and the West Country, while London, the South-East and the Eastern Counties lined up behind the Parliamentarian army whose steel helmets led to their nickname: the Roundheads.

Staffordshire in the Midlands was something of a buffer state. Allegiance was roughly evenly divided, with some towns and great houses garrisoned by Royalists and others by Parliamentarians. The county was of huge strategic significance to whichever side could wrest control.

The Civil War came to Lichfield when the Royalist Earl of Chesterfield arrived in the town with a small force of thirty dragoons just before Christmas 1642. In the coming months local noblemen brought their own troops, and by March 1643 the Lichfield garrison had grown to about 300 men. The most obvious place to station them was in The Close, alongside the cathedral, where they could take advantage of its ancient defences.

Chesterfield's Royalist force was not wholly impressive, however. The 300 men inside The Close were badly trained, poorly equipped, inadequately provisioned and, by all accounts, ill led. Chesterfield was fifty-nine, old for a fighting man in any age, venerable in the seventeenth century. He was also an ineffectual commander, whatever ability he possessed hampered by chronic gout.

Robert Greville, the second Lord Brooke, on the other hand, was a 35-year-old charismatic leader from an influential and distinguished family in the neighbouring county of Warwickshire. He had become a Member of Parliament when he was just nineteen; later in life his adventurous nature had taken him into one of the early trading companies that founded settlements in New England. He was nonetheless a strict Puritan, and as such opposed the Catholic king on religious grounds. Just before the start of the Civil War he had formed a garrison at Warwick Castle, and when war broke out he pledged his allegiance to Cromwell and Parliament. After being appointed Lord Lieutenant of Militia for Warwickshire and Staffordshire he was commissioned as a colonel in the Parliamentarian army and raised a regiment of foot in London numbering a thousand men at the outbreak of the Civil War.

It was these men whom Brooke had marched out of London, resplendent in their distinctive purple livery, to join the Parliamentary commander, the Earl of Essex, at Edgehill, in the first great encounter of the war. Charles I, with an army raised in Wales, Yorkshire and the Midlands, had advanced towards the capital. The king's men left Shrewsbury on 12 October 1642, while the Parliamentarian forces, led by Essex, raced to intercept them. The Royalists won the race and the Parliamentarians were forced to join battle at Edgehill in southern Warwickshire on 23 October as they attempted to deny the king's progress. In a textbook battle of its time, the two evenly matched armies – each of around 13,000 men – lined up in the same formation: infantry to the centre, cavalry on the flanks. There were some musketeers but they did not play a significant part in the encounter. The king held the high ground, which made it difficult for Essex's artillery, firing uphill, to have much impact. However the slope of the hill was such that the Royalist guns had little effect. After inconclusive cavalry charges Edgehill degenerated into an infantry battle, a brutal, bruising clash of two static bodies of soldiers. The fighting was fierce and bloody, initially at pike's length – some sixteen feet – but when the armies closed and the pikes were ineffective it was hand-to-hand. They fought hour after hour like some giant murderous rugby scrum, a few feet of ground gained or lost, the exertions of the two sides turning their sweat to steam, which hung above this great seething mass on the cold October day. It was only sheer exhaustion and the fall of night that stopped the fighting.

At the end of the day there were about 3,000 dead, many of them wounded who had fallen victim to the cold in the night. So keenly matched were the two armies that their casualties were almost identical:

1,500 each. The battle, for all its carnage, proved utterly inconclusive. Although the king's troops now had cleared the road to London, they were in no state to make the onward journey. Similarly Essex's force, already exhausted by its long march to the battlefield, withdrew to Brooke's garrison at Warwick Castle to regroup.

Brooke's regiment had paid a heavy price. They had been mangled in the fight and reduced to fewer than 500 men. Even so, after Edgehill Brooke used what was left of his force to secure Warwickshire for Cromwell and ultimately captured Stratford-on-Avon. When news of the Royalist garrisoning of Lichfield reached the Earl of Essex, he ordered Brooke to march on the town. Brooke had his own battle-hardened troops with him and he was allowed 700 reinforcements from throughout Warwickshire. In addition he was given several pieces of artillery to support his cause. On the evening of 1 March 1643 they camped in a field just to the south of Lichfield, which is known today as Cromwell's Meadow. Before bedding down for the night Brooke led his 1,200 troops in prayer and asked for a sign from the Almighty that right was on their side.

Although strict and stern, Brooke was a hugely influential leader, and the Earl of Essex must have felt secure in the knowledge that the 'grim-faced Lord Brooke' would bring the cathedral city into Parliamentarian control. Brooke himself felt confident as he drew up plans to order his artillery into the fray. Seventeenth-century artillery was used more for psychological effect than destructive power, but he would doubtless have believed that he could soon begin to pound the cathedral defences into submission.

The siege of Lichfield began on 2 March 1643, which was the feast of St Chad, the patron saint of the city. Not quite the sign from above that Brooke had been looking for, but perhaps an omen nonetheless: like the flattening of the Royal standard, it turned out to be far from propitious.

The day started well for Brooke. As the Parliamentarians poured into the town early in the morning of 2 March it seemed as though they would swiftly carry the day. As the streets and alleyways filled with smoke from muskets and the air rang with the clash of swords and the anguished cries of the maimed and dying, Brooke's men began to gain the upper hand. The Cavaliers fought a rearguard action through the narrow winding streets until they could seek the protection of The Close. Brooke meanwhile was able to set up a command post in Sadler Street, the modern-day Market Street, barely half a mile from the cathedral itself.

Brooke's strong Puritan convictions required the destruction of the towering Gothic cathedral topped by the king's standard, symbol of everything Brooke had dedicated his life to defeating. The Parliamentarians did not have a traditional siege cannon, but they did have a demi-culverin. It fired a ball weighing about ten pounds – smaller than the traditional siege cannons, but with the advantage of a longer range. Its maximum range was around 5,000 yards in perfect conditions, but realistically it was effective up to about 800 yards. Certainly Brooke's troops believed it would do the trick. They had given it a nickname

– Black Bess – suggesting affection and more than sufficient faith in its abilities. Supporting Black Bess, Brooke's force also had a number of smaller artillery pieces. He decided to begin his attack with a bombardment of the South Gate of The Close from Dam Street. Black Bess was ordered into the line and sited around 100 yards away.

The Royalists were not idle spectators as the Parliamentarians made their preparations. They continued to pour musket fire into the attackers as they set up their field artillery. The defenders of The Close knew that Black Bess would effectively be operating at point-blank range. It was imperative that the gun crew be put out of action as quickly as possible. The Roundheads' initial artillery barrage did little more than punch holes in the wooden gates and blow large pieces of masonry off the walls, but it was obvious that in time they would do enough to allow Brooke and his troops to pour into The Close. The Royalists, heavily outnumbered, could expect to be slaughtered by the invading Parliamentarians and the city would be lost.

Two things happened next in fairly quick succession.

On the Royalist side we know that two men, growing increasingly frustrated by their inability to inflict any damage on the gunners, decided to seek another solution. They left the gatehouse walls and climbed on to the roof of the central spire of the cathedral. This was a journey that many of the defenders would have made, because lead from the cathedral roof was being melted down to make musket balls to supplement ammunition. The central spire is some

250 feet tall and they must also have hoped that from there they might have a good enough view to perhaps pick off some of the gun crew as they went about their work. Time was against them as they clambered along the slopes of the spire, risking being shot at themselves.

On the Parliamentarian side we know that Brooke too was becoming frustrated. Black Bess was not having the instant effect he had hoped for and he decided to move forward to assess the situation personally. Brooke left his command centre in Sadler Street and made his way to the back of a house in Dam Street. As he stepped out from the doorway to move towards the gun a shot rang out from the cathedral. Brooke fell dead. He was the first recorded British victim of a sniper.

There are two schools of thought as to how Brooke met his end. The first is simply that he had the great misfortune to be in the wrong place at the wrong time, the random victim of a stray shot. Some even suggest that he was sitting by a window watching the battle when he was struck by a ricochet. The supporters of this theory offer no concrete evidence to back it up, on the basis that it requires none: it was entirely the product of chance.

The second opinion suggests that Robert Greville, the second Lord Brooke, was indeed the victim of a deliberate sniper attack.

One of the two men who had left the South Gate to go up on to the cathedral roof was John Dyott, the brother of Richard Dyott, an influential local figure and Royalist supporter. He was armed with a fowling piece, a gun normally used for shooting ducks. About six feet long with a barrel about an inch and a half in diameter, it was a formidable weapon, especially when loaded with a ball that had been made from the lead taken from the cathedral roof and melted for that express purpose. The 'bullet' would have been only a little smaller than a golf ball and much, much heavier.

John Dyott and his comrade in arms had clambered on to the cathedral roof with the express purpose of trying to do whatever they could to turn the moment to their advantage. It is argued that Dyott and the other man were on the roof when they saw Brooke watching his gun crew from the doorway. Brooke had been shouting at his gunners but, again frustrated, he stepped out from the protection of the door lintel to give more detailed instruction. As he did so John Dyott, who had been watching intently, took careful aim and fired. Brooke was wearing a purple tunic with full breastplate, as well as a steel helmet with a barred visor – the uniform is currently on display at Warwick Castle – but none of this protected him from the fatal shot. The musket ball struck him in the left eye and he died instantly.

An unnamed witness gave a graphic account of the death of the Parliamentarian commander. According to this version, Brooke had made his way towards the artillery crew after rejecting an offer of surrender from Chesterfield.

. . . my Lord would not consent to, nor did the commanders think fit, so they set to it again. My Lord observing something that he liked not in the fight, put forth his head out of the window to direct, which he had no sooner done but unhappily a bullet stroke [struck] him in the left eye, which instantly put an end to his life without speaking one word. This much enraged us and put this resolution into us, that we would die every man, but we would take the place and put every man to the sword in it.

The weight of available evidence tends to support the belief that it was a deliberate action that led to the death of Lord Brooke. However, some doubts remain. If you stand in that same doorway, underneath the plaque which marks the spot where Brooke fell, the distance to the roof of the cathedral is considerable, perhaps as much as 200 yards. Could a man really hope to take aim and hit a target with such a rudimentary weapon at that range? To make a fatal shot, from such a precarious vantage point, in the heat of battle would be a tidy piece of marksmanship even with sophisticated weaponry. To make a similar shot 350 years ago with a cannibalised gun normally used for shooting ducks is nothing short of remarkable.

Clarendon's official account of the Civil War suggests that Brooke was killed as the result of a shot fired by 'a common soldier'. John Dyott, it has been argued, as the brother of such an influential Royalist, would hardly have fitted that description. However, John Dyott was deaf and mute from birth, known locally as 'Dumb Dyott'. As such, no matter how much patronage he received from his famous brother he would not have been able to discharge the duties of an officer and is likely to have remained a 'common soldier'.

Historians may still debate the point, but in Lichfield itself you would find it hard to convince anyone that Lord Brooke died at the hands of anyone other than Dumb Dyott. The Dyott family certainly take credit for the kill, and the deadly gun is allegedly still in the family's possession today.

The death of Lord Brooke had an instant effect on his men. They poured forward with renewed aggression, spurred on by their desire to avenge their much-loved commander. While the Roundheads sought a replacement for Brooke, his men were led by his second-in-command, Sir John Gell. He played his part in inciting the attackers by putting a bounty on the Earl of Chesterfield, dead or alive, as well as offering a reward of 'fortie shillings a man' to the first ten soldiers to enter The Close after its walls had been breached.

Eventually the Parliamentarians brought a large mortar into play, which led to another attempt by Chesterfield to sue for peace. The defenders of The Close tried to arrange a parley with Brooke, unaware that he was dead. Chesterfield's terms were rejected and harsher ones dictated by Sir John Gell and the other surviving officers. Gell's conditions were accepted after some consideration, and Chesterfield and his men finally surrendered in return for not being put to death.

The death of Lord Brooke was a serious blow to the Parliamentary cause in the Midlands. It left a significant gap in the leadership: men of his presence and acumen were rare. Perhaps more importantly, the Parliamentarians missed his influence in a relatively turbulent area. As a well-known and energetic supporter of the cause, Brooke was skilled at forging alliances and rallying supporters to Cromwell's banner.

No one can certify how Lord Brooke died, other than to say that it was from a musket ball to the head. The stray shot theory has its supporters, but the notion of an aimed shot has more romance. It pits the single marksman against his victim. It personalises the otherwise random violence of the battlefield. It provides a narrative amid the chaos, a kind of brutal poetry: one shot, one kill.

The prominent victim whose death carries disproportionate influence, a blow to morale and organisation, will be remembered heroically. But the shooter? He melts away into the mists of history and uncertainty, if he ever existed at all. That pattern will be repeated again and again for hundreds of years as the sniper perfects his shadowy craft.

1

SLINGS AND ARROWS

In the opening sequence of *2001: A Space Odyssey* director Stanley Kubrick and his collaborator, the scientist and novelist Arthur C. Clarke, suggest that humankind's first evolutionary step was not to use tools constructively but to convert them to weapons. Whether it was a rock or a stone, or the jawbone of an ass, if it could be gripped and swung then it could be used as a weapon. Loud screams and fierce looks might intimidate opponents but weapons settle the argument.

With the development of human intelligence came the parallel development of weaponry. We discovered for example that certain types of rock would splinter along natural fault lines to leave an edge, then that this edge could be improved by striking it against a harder surface. Fixed to a handle that stone becomes a weapon that will easily split the skull of an enemy.

Fundamental discoveries such as fire and cooking happened largely by accident. So too did our progress in this prehistoric arms race. The man who noticed that branches carbonised by the fire after the tree was struck by lightning were harder and sharper than green branches that might bend or splinter probably became the alpha male in his tribe of hunter-gatherers.

These discoveries had all been made by about 30,000 BC. But all of these sharp stakes and axes and clubs had to be used within arm's reach of an opponent, in risky proximity. If your enemy was more skilful, or simply had a longer reach, then the advantage quickly evaporated. What primitive mankind sought was a weapon that could be used from a distance, leaving its user exposed to minimal risk of injury. The javelin and its forerunners were fine, since they could be hurled a considerable distance and with reasonable accuracy. They tended however only to be really effective when thrown en masse. These early javelins were not the sleek, aerodynamically designed models of modern athletics events; they had barbs at one end so that the point would stick in the wound. Later they were designed so that the shaft would break away from the head on impact, so that it could not be recovered and hurled back. But the javelin was large and cumbersome, and limiting in the numbers that could be easily carried into a fight.

Two technologies offered the advantage of being able to attack a distant target repeatedly: the sling and the bow appear in human conflict for the first time, and they represent an evolutionary leap in weaponry. They are primitive missile systems. Both could be devastating in their accuracy and both could be used without exposing the user to too much risk.

The sling is the simplest of weapons; few are more elegantly efficient. Take a good round stone about the size of a man's fist, place it in the centre of a long narrow strip of cloth or hide, grab both ends of the strip together and then whirl it around your head as fast as you are able. Releasing one end of the strip produces a combination of centrifugal force and potential energy, which will turn your rock into a missile capable of travelling a considerable distance with a good degree of accuracy. Slingstones travel too fast to be seen or dodged; used en masse they can have a devastating effect.

The bow and arrow is similarly remarkable as an elegant marriage of form and function. All it took was a branch bent into a 'D', the gut of a dead animal strung between the two end points, and a sharpened twig or branch notched at one end so it could be held on the bowstring. With practice the bow became man's most useful tool in the hunt. No longer was it necessary to wait for an animal to stray into the noose of a carefully set snare. No longer would the hunter howl in frustration as a beast startled by a noise or the scent of his presence bolted away while a spear was still in flight. Instead an arrow could be delivered with sufficient speed and force to kill an animal almost before it had been alerted by the noise of the bowstring. And if the first shot missed then arrows were light enough to carry in abundance, giving the chance of successive follow-up shots.

Although probably developed as hunting tools, slings and arrows were easily and inevitably adapted as weapons of war. The most famous slingshot of all is recounted in the Old Testament in the Book of Samuel, where the Israelite shepherd boy David slays the Philistine giant Goliath with a single shot from

17

the sling that he had used to protect his sheep from predators.

The most significant development in adapting slings and bows for warlike purposes came after the discovery of the properties of copper around 8,000 years ago. Once it was realised that copper and tin could be smelted together to form bronze, then bows could be made from metal or from wood/metal composites, vastly increasing their range and power. Similarly, wooden arrows could be tipped with metal points to inflict even more damage. Around 1000 BC, man discovered how to smelt iron ore, and this added a new and effective range of materials to the arsenal.

The first battle for which substantial records survive, though by no means the first battle ever to be fought, was at Megiddo, a pass not far from what we now know as Mount Carmel in Israel. It took place in the early fifteenth century BC and it was fought by the army of Pharaoh Thutmoses III against an alliance of rebel tribes. It was a significant victory for the young pharaoh, since it enabled him to increase Egypt's dominance from Syria in the north to the River Euphrates in the east. Taking a longer view, the battle at Megiddo is important because it is the first recorded use of archery in war. Thutmoses' army attacked in a crescent formation with swords and spears for close-in fighting, supported by massed ranks of bowmen who fired into the enemy formation as they advanced.

The military use of the archer coincided with the development of the chariot. In the second millennium BC, Egypt under leaders such as Thutmoses III and Ramesses II achieved military dominance in the

18

region because of their use of chariots, which provided a mobile firing platform for an archer. The early chariots were light, often made of wicker, and large enough only for the archer and a driver, who might wear protective armour. The chariots were pulled by a pair of horses and, although they were limited in their manoeuvrability by their fixed wheels and wide turning circles, an attack by massed charioteers charging into an enemy that had been demoralised by flight after flight of arrows loosed into their ranks made the Egyptians all but unbeatable.

Other races inevitably developed chariot technology of their own, foremost among them the Hittites, who were Egypt's main rivals in the region, and the Assyrians. In the seventh century BC, when Assyria effectively dedicated itself to being a military power, the whole state was built around a professional army. This army was self-financing. It would conquer neighbouring territories, and the economy of the Assyrian dominion depended almost entirely on what was looted from conquered neighbours. Using new technology – most of their weapons were made of iron – the Assyrians employed sophisticated tactics based around the bow and the chariot to establish military superiority. The Assyrian archers would unleash wave upon wave of metal-tipped arrows into the ranks of the enemy before their iron-clad chariots would smash through the opposing infantry.

Some of the best slingers in the known ancient world came from Greece, especially Rhodes. The Rhodian slingers were capable of outdistancing archers with well-aimed shots estimated at around 350 metres.

Their missiles were no longer confined to smooth stones. They could be of either ceramic or lead. There are examples of lead slingstones being found with slogans carved on their soft surface, much as Second World War bomb crews would chalk messages on the bombs they were about to drop. The acorn-shaped slingstone could be doubly effective. A direct hit would frequently be delivered with enough force to be lethal, but even a wound could eventually prove fatal. The stone could break the skin and lodge in the flesh, where its smooth surface made it difficult to remove. If the wound closed over and infection set in, the victim was doomed to a lingering and intensely painful death.

Philip of Macedon used both slingers and bowmen in the army he bequeathed to his son, Alexander the Great, who used it to build the greatest empire the world had seen. The Romans too used slingshots, and each legionary was supposed to carry one. In later years a pole-mounted sling was developed which would deliver a larger payload and can be seen as a forerunner of the siege machine.

It was a force of slingers that helped inflict on Rome one of its heaviest defeats. In the Second Punic War, the Romans faced the Carthaginians under the command of Hannibal. Hannibal's army was largely composed of mercenaries from Spain, France and North Africa; it was confronted by the supreme discipline of the Roman legions. But at Cannae in Southern Italy in 216 BC Hannibal won the day. His slingers were from the Balearics and, noted for their combination of skill and savagery, were the most useful part of Hannibal's infantry. They carried three slings of differing lengths which allowed them to commence fire at long range, continue to harass the enemy as they closed, and deliver a final deadly volley at short range before the two masses of infantry collided. Although they would normally be supported by the war elephants with which the Carthaginian general is for ever associated, these massive weapons were all dead by the time his army reached Cannae. The slingers operated without support that day, but were able to act with such precision at such a distance that they caused huge damage to the Roman ranks. His cavalry finally won the day for Hannibal at Cannae, but his success was built on the use of the slingers.

The Roman legion remained the region's predominant force even after Cannae. To oppose it required luck and ingenuity. It isn't surprising, therefore, that a significant development in the history of weaponry occurred at the scene of another Roman defeat.

Almost 150 years after Cannae Rome had begun to look toward the East for further conquests. The wealthy aristocrat Publius Crassus gambled on an invasion of Parthia, just to the east of modern Syria, to win him enough popular support to advance his political ambitions over Julius Caesar. Acting on his own authority he invaded Parthia in 53 BC, but his army was intercepted at Carrhae by a Parthian force commanded by Surena, a ruthless but brilliant general. Surena had the initial advantage of having chosen his ground. The Roman legions found themselves surrounded on a flat plain, far from their base, and without cavalry support. Surena's force was about the same size as Crassus's, but included some 10,000 cavalry, all using bows, with a supply train of 1,000 camels whose sole function was to resupply the archers with arrows.

The bows used at Carrhae were vastly different from the crude devices favoured by early hunters. The basic wooden bow had given way to a composite construction often of wood and horn. Wood was still the starting point, but it was now embellished and augmented. Thin layers or strips of horn would be glued to the belly of the bow, that part which faces the archer, to give it strength, while on the other side layers of animal sinew were stuck on. Two conflicting forces – the compression of the horn and the elasticity of the sinew – produced a weapon that would unleash a considerable amount of power.

The biggest limitation of horse-mounted cavalry was that it was hard to fire a bow from horseback with any degree of accuracy while still controlling the horse as you raced towards your enemy. The development of the saddle and stirrups, probably in India in the first century BC, changed all of that.

The Parthians had quickly mastered this new technology, developing what came to be known as 'the Parthian shot'. Surena's cavalry could race full-tilt firing arrows from horseback at the Roman legions. The Romans' battle tactics relied heavily on the protection of their shields; their normal defensive position was 'the testudo' – literally 'the tortoise' – in which the shields were held over their heads and at their sides to form a protective wall of armour that would deflect most enemy fire. The testudo could then move forward and batter its way through opposing lines with bone-breaking force. This however is only effective if you have an enemy who is willing to engage a lumbering, essentially static force with one of its own. But rather than meet the Romans head-on, the Parthians could wheel away at the last moment and let fly. The Parthian shot was simply the ability of the Parthian archers to turn in their saddles and fire accurately behind them as they were racing back to their own lines.

By using these hit-and-run tactics, Surena systematically picked Crassus's legions apart. Shower after shower of arrows fell on the Roman shields, eventually forcing them to retreat. But there was nowhere to go. Small clusters of men would be separated from the main body and slaughtered without mercy. The Romans were helpless. In the right circumstances their short stabbing sword – the gladius – was a fearsome weapon, but it required the enemy to be no more than five feet away to be effective. Here, with an enemy that refused to engage on those terms, they were trapped by the size of their force and their lack of mobility. Crassus, facing defeat and disgrace, attempted to negotiate his way out, but Surena had him executed and his troops put to the sword.

After Carrhae, the horse-mounted bowman was all but invincible on the battlefield until European armies developed effective counter-measures, principally the crossbow, almost a thousand years later. It had previously been difficult for a bareback rider, or one using merely a horse blanket for comfort, to fire a bow in combat. Now, whether using the Parthian shot or not, the limitations on the horse-bowman had been removed. Within 500 years, Attila the Hun with an army composed almost entirely of horse-mounted bowmen would sweep out of Asia through Southern Europe and to the gates of Rome itself. Every one of Attila's men went into battle on horseback with a bow and several quivers full of arrows. In addition he could have as many as half a dozen horses in reserve, each of them as heavily supplied as the other. Hun archers could hit a man at a distance of 100 metres on a flat trajectory, double that if they were firing into a massed enemy where they would unleash their arrows in a high, steepling, parabolic trajectory.

By the fifth century AD weapons development was racing far ahead of battlefield tactics. Combat still took place more or less by appointment, with two opposing groups rushing at each other until one was crushed into submission or both ground themselves to a stalemate. Even the Roman army, for all its vaunted sophistication in engineering, still fought as a phalanx, battering its way into the enemy before cutting them to pieces at close range with the gladius. However, what military strategists might later recognise as the 'weapon of decision' was beginning to emerge.

The effectiveness of a weapon was beginning to be influenced as much by the user as the weapon itself. Employed correctly it could decide the outcome of the battle. The archer had been a remote combatant who would unleash volleys of arrows in great clouds towards the enemy from relative safety. As such he was frequently scorned by infantrymen. Once a bow-man was put on horseback he could carry the fight to the enemy, picking his shots, but at the same time exposing himself to danger. This shift towards personal skill rather than massed fire would be emphasised even more in another major development.

The crossbow was a simple but deadly invention. Generally it comprised a wooden stock or handle with a bow made of wood or metal lying across it at right angles. The thick bowstring is too heavy to draw back manually, but it can be winched back, or spanned, by levers and held in place by a trigger and lock mechanism. A bolt, or quarrel, can then be placed in a groove that runs the length of the top of the stock. When the trigger is released the bowstring whips forward at considerable speed, catching the end of the bolt and propelling it towards the target with great force and accuracy. As with the conventional bow, composite forms also existed, with horn or whalebone placed between two pieces of wood for added power.

The crossbow, or arbalest as it is also sometimes known, seems to have been developed in China some time during the fourth century BC. The first mention of its existence is in an account of the Battle of Ma-Ling in 341 BC. Later still, in the first century AD, the historian Heron of Alexandria describes the use of a device he refers to as a gastraphetes or 'belly bow', which was spanned by placing the stock against the stomach and pushing a slide forward to lock with a trigger. The Romans, with their ability to cherry-pick the best of the technology from the lands they had conquered, brought the crossbow to Europe. The historian Vegetius Romanus wrote in AD 385 in *De Re Militari – About the Military* – that crossbowmen were an accepted part of the Roman army's establishment. Some historians suggest that the matter-of-fact manner in which these crossbowmen were mentioned by Vegetius indicates that they were far from a novelty.

But with the fall of the Roman Empire, the crossbow slowly disappeared from Europe. It would be a further 500 years before it was 'reinvented' by the Normans, drawing on huge technological advances in the intervening half-millennium. The materials from which the crossbows were made had improved radically. The bows themselves were much stronger – the steel bow would appear by the fourteenth century – and methods of firing them had also improved.

The ammunition had become more deadly too. No longer content with arrows or wooden bolts, the crossbowman now fired a quarrel, a hardwood bolt tipped with a tempered metal point. A quarrel could reliably penetrate a shield or armour at a distance of up to 150 metres. The crossbow was lighter and its user could be more mobile on the battlefield. Although crossbowmen often had the freedom of the battlefield to select their targets, they were not, by a long way, the first snipers. But with weapons that could be aimed and fired in a more or less flat trajectory, a case can be made for these men as the earliest sharpshooters.

The crossbow had its drawbacks. Bulkier than a longbow, it also took longer to fire. Its range was limited by the physics of its own construction too: the weight of the cord required to propel the quarrel meant that it absorbed a lot of the energy released when the weapon was fired. A crossbow was generally effective only up to about 200 yards. Nonetheless in the eleventh century it was gaining a reputation as the ultimate weapon, since even at the extremes of its range it could puncture chain mail or armour and inflict a devastating and often fatal wound.

The crossbow also threatened the social order of the day. By the time it made its reappearance on the battlefields of Europe in the eleventh century, war had become the ultimate manifestation of the feudal system. Only the nobility were afforded protection in the form of suits of armour, while the serfs and mercenaries made do with chain mail or leather. Generally speaking, the only person who could inflict damage on a fully armoured nobleman would be one of his peers. A crossbow in the wrong hands, however, became potentially an instrument for social change by allowing a serf to deal a mortal wound to a feudal lord, and at a considerable distance. So effective had the crossbow become in penetrating armour that crossbowmen would frequently shoot the horses out from under the armoured knights rather than killing their riders. This allowed the knights to be captured as a valuable source of ransom.

A nobleman could find himself unhorsed, captured, and his lands taken by a rival in his absence with the whip of a crossbow string. Where once the armoured knight had been as invulnerable as Achilles, now, like him, he could be struck down as if by lightning. The crossbowman became so valuable that many were able to make a lucrative living as mercenaries.

The crossbow so threatened the social status quo that the Church was eventually minded to intervene. At the Second Lateran Council in 1139, Pope Innocent II forbade its use under penalty of anathema, or excommunication. The crossbow was described by the Pontiff as a weapon that was 'hateful to God and unfit for Christians'. Similarly, in England Magna Carta forbade the employment of foreign mercenary crossbowmen. Despite the proscriptions of the Church the crossbow – the most powerful missile system of its time – was still widely adopted throughout Europe, with the exception of England.

The Lateran Council decree is the first recorded attempt at arms control, a SALT II of its day. However like most arms-limitation treaties the wording of the ruling was carefully couched so that it left a significant loophole. Crossbows could not be used against Christians, but there was no reason why they could not be used against 'infidels', which meant that Christian knights were perfectly free to use crossbowmen in their battles during the Crusades, launched by Pope Urban II in 1095 in Clermont in France.

It was the Crusades that made the military reputation of one of England's kings, the crossbow that was both his weapon of choice and his nemesis. King Richard I of England came to the throne in September 1189. Nicknamed by his men Coeur de Lion – Lion Heart – for his valour in battle, he was king of England in name only. Although English-born, he preferred the attractions of his Plantagenet home in France to England, in which he spent barely six months of his ten-year reign. In any event he had no sooner taken the throne than he left to fulfil a promise to his late father Henry II that he would join the Third Crusade.

Richard left England in 1190, taking with him

an army of some 55 knights and 2,000 infantry. However he had used the Lateran loophole wisely, and since the business at hand was slaughtering infidels most of his infantry, certainly more than 1,000, were crossbowmen. The crossbow was a particularly effective weapon in the siege conditions the Crusaders expected to face, where a premium was placed on accuracy rather than rate of fire.

Richard is now considered to be one of the two greatest generals of his time. The other was his rival in Palestine, Saladin. The Lion Heart's grasp of tactics brought him some notable victories. At Arsuf in September 1191 Richard, who was marching from Accra to Jaffa, met the armies of Saladin. With the sea protecting one flank, Richard was able to advance with his crossbowmen protecting the other. This meant that the cavalry in the middle of the force were able to concentrate all their efforts on the Muslim army. Saladin ordered three charges, each with ferocious force, and each was repelled by Richard's army.

Despite his successes in the Crusades, Richard never took Jerusalem. Twice he came close, but in the end he realised that he had enough men to take the city, but not to hold it. A truce was brokered with Saladin that allowed him to return to Europe. Famously he was kidnapped and ransomed en route by Leopold of Austria.

The man who had fought the holiest fight died as a result of a much pettier squabble in his mother's native Aquitaine. A peasant digging in a field had found a treasure in the village of Châlus in Limousin and, as was his right, his liege lord had claimed it as his own. Richard then exercised his own right to claim the treasure from the nobleman. When the nobleman refused, Richard and his army laid siege to Châlus. Perhaps believing that the great Coeur de Lion was somehow invincible, Richard did not wear full armour during the siege. One day as he was riding near the castle walls, a crossbowman stood up on the parapet to aim a shot at him. Some sources say that Richard was so taken with the man's audacity that he stopped his horse and waved to his attacker. What is not in doubt is that the crossbow bolt found its mark in Richard's shoulder. The king waved off his anxious courtiers and declined treatment for what he believed to be a minor wound. Tragically it became infected and ultimately gangrenous. Richard died on 6 April 1199 at the age of forty-one.

A single well-aimed shot had caused the death of a king. Despite the fact that the crossbow was bulky and cumbersome, this was its ultimate advantage. One man, taking careful aim against an unwary opponent, could change the course of a battle or indeed history.

Joan of Arc was also the victim of a crossbow shot, on 7 May 1429 during the siege of Orléans. She was not killed, but her wounds were so serious that her recovery was thought to be proof of the Divine Right of her cause. Notwithstanding the theological arguments, it does seem likely that since Joan was known for her white armour, she was the victim of a crossbowman who had been given specific instructions to shoot her.

The crossbowman had become a deadly and feared enemy. It is hardly surprising that any taken prisoner were frequently summarily executed or, if not, then mutilated to prevent them ever firing the weapon again. The strategic value of the crossbow would be proved repeatedly throughout the High Middle Ages. Arguably the last time – before other technologies superseded it – that it was crucially decisive was in the siege of Constantinople.

By the second half of the fifteenth century the face of warfare had changed with the introduction of gunpowder to Europe. Even so, the principles embodied by the crossbow, a manoeuvrable weapon of deadly accuracy that could pick out a target of the highest rank, would be retained and find its apotheosis in the concept of the sniper. Indeed this was demonstrated in one of the first military engagements in which gunpowder played a major part.

The Turks used artillery to deadly effect at the siege of Constantinople in 1453. They simply camped outside the city and used their heavy cannon. The walls were susceptible to bombardment and often simply collapsed under their own weight after the first few shots.

The defenders of the city were cut off and approaching desperation. Almost all of their hopes rested on a Genoese knight, Giovanni Giustiniani, who had arrived in the city in January with 700 well-armed men. The Emperor Constantine immediately named him as defender of the city. From the beginning of the year the Ottoman leader Mohammed II began to assemble a force that would eventually amount to some 150,000 men. Inside the city walls Giustiniani, terribly outnumbered, marshalled his forces brilliantly to try to frustrate Mohammed.

The Turks began their first attack on 18 April. They had been bombarding the city for days before they attempted to storm the walls. Giustiniani and his men fought bravely for four hours, often hand to hand, and because they had chosen their defensive positions wisely were able to beat back the Turks. The pattern would be repeated many times during the coming days and weeks. Finally, on 29 May, Mohammed threw his best troops, the Janissaries, into the battle.

The battle raged through the night, until almost at daybreak a shot from an unnamed Turkish soldier hit Giustiniani. Some sources say he was hit by a cross-bow quarrel, others suggest it may have been a ball from a primitive forerunner of the musket. Certainly there would have been Turks armed with crossbows, a weapon with which they were particularly proficient, and the presence of cannon at the battle means we cannot rule out the presence of crude small arms. Whichever weapon fired the shot, the effects were catastrophic for the city. Although not killed, Giustiniani was gravely wounded. The emperor begged him to stay at his post, but Giustiniani asked to be taken from the field with the other casualties. A gate was opened and he was carried out by his own men. Other defenders saw him and assumed that the battle had been lost, so they took the chance and fled through the open gate. The Janissaries, spotting the breach in the defences, poured into the city.

The Janissaries were followed by thousands of other Ottoman troops, and Constantinople was taken. With Giustiniani, their talismanic leader, incapacitated the exhausted defenders became demoralised and were able to offer only token resistance. Giustiniani himself was carried to a ship and escaped capture but died of his wounds a few days later.

Constantinople fell, and with it came the beginning of the end for the Byzantine Empire. The city would doubtless have been taken eventually, but the absence of Giovanni Giustiniani was decisive. The significance of his death is that he was a victim of a well-aimed shot from a single soldier. As would often be the case, the identity of the man who fired the decisive shot would remain a mystery. It is poignant that we cannot be sure if Giustiniani was the last significant fatality caused by the crossbow or one of the first to fall victim to a new technology that would transform the battlefield.

Gunpowder had arrived. The rules of engagement had changed for ever. The age of chivalry was dead.

2

SURE SHOT TIM

There is some uncertainty about who developed black powder – a mixture of saltpetre, sulphur and charcoal. The Chinese were using it simply for pyrotechnic displays by the end of the first millennium. It was also being used for fireworks in the Arab world before making its appearance in the West towards the middle of the thirteenth century. By the early fourteenth century, the idea of placing this incendiary powder in a tube with one end sealed so that it could propel a missile from the other end was well established.

The advent of gunpowder, as it became universally known, led to the widespread introduction of small arms from the mid-thirteenth century, but these were neither powerful nor accurate. About two hundred years later came the arrival of the arquebus, to all intents and purposes the first reliable handgun.

The arquebus had a curved stock, which meant it could be aimed. It also had a fuse or match, usually a length of cord soaked in saltpetre and allowed to dry. Smouldering after being lit, this could be touched to the powder in the priming pan of the gun, igniting the main charge and firing the shot. The arquebus, also known as the hackbut, could fire a one-ounce ball at a maximum distance of between 100 and 200 yards; the rate of fire was slow, however, and two shots in three minutes was deemed exceptional. The arquebus had many shortcomings – for example it was generally not powerful enough to penetrate armour. Nevertheless, this crude weapon became the standard for the next hundred years or so.

It was eventually replaced by the matchlock, a closer relative to the musket. Bigger and heavier than the arquebus, the matchlock could be up to seven feet long and weigh as much as 25 pounds. It fired a ball of about an ounce and a half. The extra size allowed it to hold more powder: it was more powerful. Its theoretical range was up to 600 yards, although 200 yards was more usual. The additional weight made the matchlock relatively immobile. It had to be balanced on a forked rest to aim and fire, and was effectively more of a small field piece than a musket. It may have been an adapted version of a piece such as this that John Dyott used to kill Lord Brooke.

Even when improvements in technology reduced the weight of the musket to a barely more manageable 14 pounds there were still drawbacks; the smouldering match gave away the position of the musketeer and, more significantly, the matchlock musket could only be used if it wasn't raining!

The great era of English sea power owed much to another development which ran in tandem with the introduction of gunpowder; the invention of reliable and accurate ship-mounted cannon. By the sixteenth century English colonisation stretched across the globe. The example was adopted by other European countries, and with the sturdier more durable ships which had to be developed to accommodate the cannon, what became known as the New World opened up to Europeans.

A European settler in the colonies of eastern North America in the eighteenth century needed a number of basic skills to survive. A durable shelter would have to be built, rudimentary carpentry would be needed to furnish it, and some knowledge of farming would be important to grow crops. But the most important and immediate skill for a settler to master was the ability to shoot well.

If you could not shoot you would be unable to defend your land against the Native Americans from whom you were stealing it in the first place. But above all else, if you could not shoot you would starve. Newly arrived settlers could not wait for crops to grow. They needed to hunt to put meat on the table, and not just with snares or traps. Larger game was required to survive harsher seasons. The settlers had to be able to track and shoot.

Many of the early European settlers came from Continental Europe, specifically Austria, Germany and Switzerland. These were men who had been schooled in the gun from a very early age. Shooting competitions were a highlight of town and village life in Europe, and people travelled from far and wide to demonstrate their skill. Europeans brought their guns to the New World. The traditional Jäger rifle – it was named after the German word for hunter – was modified in the Americas by local gunsmiths. They lengthened the barrel for greater accuracy and efficiency, and reduced the bore of the weapon to take a smaller musket ball, which preserved scarce raw materials. This new variant was known as the Pennsylvania rifle, since many of the migrants had settled in that state, although later it would also be known as the Kentucky rifle thanks to the proficiency of the Southern huntsmen.

The difference between a rifle and a musket lies in the barrel. The musket is essentially a smooth-barrelled weapon with limited range and accuracy; the rifle on the other hand has spiral grooves ground into the inside of the barrel. This imparts spin on to the ball as it leaves the barrel, allowing it to fly further and truer. The musket could be fired more quickly and was the weapon of the common soldier; the rifle took longer to load but was more accurate and more deadly. It was the weapon of a hunter or a marksman.

When the American Revolution began in 1775, the rifle stopped being used for hunting game and started being used to hunt British soldiers. Farmers and settlers who could drop a squirrel as it ran along a branch suddenly found their skills much in demand against Redcoats.

There is a misconception that the American War of Independence was fought between the British on the one side, resplendent in their scarlet tunics, and buckskin-clad frontiersmen on the other. These pioneers, so the story goes, would shoot from cover while the British marched line abreast, a blaze of red through the forest, with fife and drums sounding just to make sure everyone knew they were coming. A great many frontiersmen did indeed fight from the woods, but in fact the Continental American Army, with its blue uniforms every bit as conspicuous as the British scarlet, conducted itself in much the same way as its British enemy. Tactics had not changed that much since ancient times either. Engagements were fought by massed ranks of men marching towards each other until they were close enough to unleash a fusillade of volley fire. Since the smoothbore musket was largely ineffective beyond 80 yards and could be loaded only three or four times a minute, an infantry-man would be lucky to have two effective shots before the lines clashed and the battle resorted to swords and bayonets.

The British Army was equipped with the Brown Bess, a smoothbore flintlock musket that fired a ball that weighed a little over an ounce. It was robust, dependable, but not very accurate. Major George Hanger, a firearms expert who served with the British in the Revolution, was scathing in his assessment of the capabilities of the Brown Bess:

A soldier's musket, if not exceedingly ill-bored (as many of them are), will strike the figure of a man at 80 yards; it may even at 100, but a soldier must be very unfortunate who shall be wounded by a common musket at 150 yards, providing his antagonist aims at him; and as to firing at a man at 200 yards with a common musket, you may just as well fire at the moon and have the same hopes of hitting your object. I do maintain and will prove, whenever called on, that no man was ever killed at 200 yards by a common soldier's musket, by the person who aimed at him.

Other Royal Ordnance tests in 1846 found that the maximum range of a musket with a five-degree elevation was about 650 yards, with the most effective range only about 75 yards. Testing its accuracy provided even more damning results. Ten shots fired at a target 11 feet 6 inches high by 6 feet wide from a distance of 250 yards all missed; five hit the target at 150 yards. The Ordnance Board came to the conclusion that firing at anything beyond 200 yards was a waste of ammunition.

Hanger's damning verdict and the Royal Ordnance tests were endorsed by a modern test of a Brown Bess attempting to simulate 1775 conditions, in which only two out of ten shots hit a target 50 yards away. At 100 yards only one shot in ten found its mark in a man-sized target – a strike rate less effective than a Balearic slinger some two thousand years previously.

The limitations of the musket dictated the way battles were fought. Armies could march towards each other firing once they reached about 150 yards away, volley fire would commence in earnest at about 80 to 100 yards, and then they would close and fight with sword and bayonet. For all its shortcomings, the musket could prove a deadly weapon when operating within its acceptable parameters. At the Battle of Borodino in 1812, for example, Napoleon lost 30,000 of an army of 130,000 killed or wounded by musket fire while the Russians lost 40,000 out of 120,000.

Firing the musket in combat was no mean feat. The British Army procedure was a very complicated process; the Americans however had simplified it to a mere twelve actions. In drills each of these specific tasks would be carried out en masse in response to orders.

1 'Half-cock your firelock' – pull back the cock of the musket into a half-cocked position.
2 'Handle your cartridge' – take the cartridge from your pouch, bite off the end of the twist of paper and pour the black powder into the pan of the musket.
3 'Prime' – prime the musket.
4 'Shut your pans' – close the musket pan.
5 'Charge with cartridge' – put the cartridge into the barrel of the musket.
6 'Draw your rammers' – take the ramrod from its position, usually between clips beneath the barrel of the gun.
7 'Ram down your charge' – use the ramrod to force the cartridge down the barrel.
8 'Return your rammers' – put the ramrod back and bring the musket to a port arms position.
9 'Shoulder your firelock' – bring the gun to your shoulder.
10 'Poise your firelock' – steady the gun in a shooting position.
11 'Cock your firelock' – pull the musket cock, which holds the flint that will provide the spark, to its full position.
12 'Fire' – and then get ready to do it all over again!

One way of improving the killing power of the musket was to load it with more than one musket ball. Loading with a musket ball as well as several pellets of buckshot became standard practice. A musket ball had a calibre of around .75 inches while buckshot was smaller, around .30 inch. The calibre of a weapon strictly refers to the diameter of the bore of the barrel expressed in hundredths of an inch, but is generally taken to refer to the diameter of the ammunition. Therefore a .75 musket ball was about three-quarters of an inch in diameter while a piece of buckshot was about three-tenths of an inch across. This 'buck and ball' mixture could inflict damage over a wider area. During the American Revolution George Washington issued orders that the Continental Army should load its weapons with one musket ball and either four or eight pieces of buckshot depending on the size and power of the musket. He later ordered that all cartridges being made should contain buckshot as well as a musket ball. It was a simple tactic: pour as much shot as possible in the general direction of the enemy and hope that it does some damage.

Although both the British and the Continental armies were equipped with the Brown Bess, individual colonists had their own rifles. These were the weapons of the minute men, a force drawn from the best of the local militia and set up in 1774, the year before the Revolution began. The minute men were so called because they had to be ready to march or fight at a minute's notice. At Lexington, Kentucky, where the first battle of the Revolution took place, the minute men trained constantly aiming at man-sized targets set afloat in the Charles River. They took turns to fire their weapons, and when they were not shooting they would make musket cartridges. No one was excused this drill until they were capable of loading and firing fifteen times in three and three-quarter minutes. The minute men and other light infantry units were also trained in bush fighting so they could better harass the British. The minute men units were eventually absorbed into the Continental Army, where their presence as trained and battle-hardened troops would prove invaluable.

The use of the rifle gave the colonists a significant edge in the Revolutionary War. Riflemen could be used as scouts and free-ranging skirmishers who could engage the British in guerrilla firefights at some distance from their own lines. They wore their own plain clothes rather than colourful uniforms, and the British could not tell where the shots were coming from, but since the Brown Bess was hopelessly inaccurate at anything more than 80 yards this was largely academic. The colonists were able to shoot with deadly accuracy at distances in excess of 300 yards while the British could only take cover. George Hanger, the man so scathing of the Brown Bess's abilities, had a close call of his own. Hanger was standing near a mill talking to Colonel Banastre Tarleton, the infamous British officer demonised by Jason Isaacs in the Mel Gibson film *The Patriot*, when as he recalls they were spotted by an American sharpshooter.

. . . [he] laid himself down on his belly; for in such positions they always lie, to take a good shot at long distance. He took a deliberate and cool shot at my friend, at me and at our bugle-horn man. Now observe how well this man shot. It was in the month of August and not a breath of wind was stirring. Colonel Tarleton's horse and mine, I am certain, were not anything more than two feet apart ... A rifle-ball passed between him and me; looking directly at the mill I evidently observed the flash of the powder ... the bugle-horn man behind me, and directly central, jumped off his horse and said 'Sir, my horse is shot'. The horse fell down and died ... Now speaking of this rifleman's shooting nothing could be better ... I have passed several times over this ground and ever observed it with the greatest attention; and I can positively assert that the distance he fired from at us was full 400 yards.

Although this shot seemed remarkable to both Hanger and, presumably, Tarleton, it was not exceptional by American standards. General George Washington was well aware of the skill of his riflemen, and one of his first acts as commander of the Continental Army was to set up what amounted to elite rifle companies. These would be 'chosen men, selected from the army at large, well acquainted with the use of rifles, and with that mode of fighting which is necessary to make them a good counterpoise to the Indian'. At Washington's request, the Continental Congress set up ten of these rifle companies in Virginia, Pennsylvania and Maryland.

These were hard men: skilled woodsmen, crack shots, and well used to the hardships of frontier life. The most famous of these riflemen was Daniel Morgan of Pennsylvania, whose exploits had already become the stuff of contemporary legend. He had marched his men 600 miles in twenty-one days to join the siege of Boston. Then he joined General Benedict Arnold on an epic march to Quebec in a failed attempt to capture the city. Morgan and his men came within a hairbreadth of taking the city, and Canada with it.

George Washington eventually commissioned Morgan to pick the 500 best shots in the army. These men were sworn in to the Continental Rifle Corps, or as it became better known, Morgan's Rifles. Only the very best were invited to join. One of those who passed the test was Timothy Murphy of Pennsylvania, the son of Irish immigrants from Donegal. Murphy was a dead shot capable of hitting a 7-inch target at 250 yards, and he had also taken part in the memorable march with the rifle regiments at the siege of Boston. He was transferred to Morgan's Rifles in July 1777 and entered the history books in its first major action.

Washington had intended to use Morgan's regiment as a light infantry force to skirmish and harass the British. Whatever his intentions, the decision was taken out of his hands. The British under General Burgoyne were threatening the Colonial troops in New York State, and Morgan and his men were dispatched as reinforcements. Burgoyne was using Indian scouts to great effect and Washington felt that Morgan's men would be the ideal opponents. Morgan's men turned out to be more than a match for Burgoyne's scouts, especially encouraged by the stirring exhortation of Benedict Arnold.

'Colonel Morgan,' Arnold told him, 'you and I have seen too many redskins to be deceived by that garb of paint and feathers; they are asses in lions' skins, Canadians and Tories; let your riflemen cure them of their borrowed plumes.'

Morgan's victory was total. Eyewitness accounts say it took less than fifteen minutes for Burgoyne's scouts to be sent running back to their own lines. Without his scouts Burgoyne's force was effectively blind, and he had no option but to move his men towards Saratoga, where he could perhaps make a stand. On 19 September 1778 the Americans intercepted three columns of Burgoyne's men who were attempting to seize control of the high ground. The two sides met in a clearing called Freeman's Farm, where Morgan and Arnold quickly gained the upper hand. Thanks to their marksmanship and the superior range of their weapons Morgan's Rifles kept Burgoyne's men pinned down in the woods. Any Redcoat who showed himself was liable to be cut down. Morgan's men attempted a charge, but when they were beaten back they simply merged into the forest and regrouped, unseen by the British. Burgoyne then tried a counter-attack but the colonists, inspired by Morgan and Arnold, held their line. At the end of the day what became known as the First Battle of Saratoga had ended with 566 British dead and 313 dead on the American side.

It says a lot about the way war was waged at the time that the two armies then rested barely a mile from each other. They spent three weeks waiting for reinforcements, with only the odd sniper patrol going out from time to time to remind them they were still at war.

On 7 October, the Second Battle of Saratoga took place when Burgoyne made an attempt to break out.

He left 800 men in his camp and advanced with a force of some 1,700 men. As he slowly made his way forward he found himself surrounded by Morgan's men. Burgoyne tried to send a message to tell his troops to fall back but the messenger was shot dead by one of Morgan's marksmen. Facing the enemy on all sides, the British began to retreat until they were rallied by General Simon Fraser.

Brigadier General Simon Fraser was a brave and courageous officer who was admired on both sides for his valour and skill. He had distinguished himself at the First Battle of Saratoga, frequently putting himself in harm's way to try to retrieve a situation that seemed lost. Burgoyne's tactics were totally ill suited to the rough terrain in which they were fighting. Nonetheless Fraser continued to follow his commander's orders, often at considerable risk. The British faced another defeat at the Second Battle of Saratoga, with Fraser, who was held in great regard by his men, perhaps the one man who could restore British fortunes.

On his distinctive grey charger, Fraser was instantly recognisable as he rode across the battlefield exhorting his men to one last great effort. When Benedict Arnold spotted him, he turned to Daniel Morgan and said: 'That man on the grey horse is a host in himself and must be disposed of.'

Morgan knew instantly what had to be done, and he called Timothy Murphy over to him.

'That gallant officer is General Fraser,' he said. 'I admire him, but it is necessary that he should die. Do your duty.'

Murphy responded instantly. He climbed a nearby

tree to get a better view of the battlefield and, finding a suitable branch, he rested his rifle in a fork in the branch. Fraser was about 300 yards away; Murphy took careful aim and fired. The first shot missed Fraser but cut one of the straps holding his saddle in place. Now that he had the range Murphy fired again. This time the bullet passed through the horse's mane, narrowly missing its head.

On the British lines it was evident that Fraser was being deliberately targeted by the Americans. One of his aides rushed towards him, urging him to retreat to safety.

'My duty forbids me to fly from danger,' responded Fraser.

As he spoke, Timothy Murphy lined up a third shot. This time he did not miss. The musket ball flew straight and true and tore into Simon Fraser's stomach. As he fell from his saddle mortally wounded, fellow officers rushed to help. Timothy Murphy fired once more and critically wounded Sir Francis Clerke, Burgoyne's aide de camp. The shooting of General Fraser came at roughly the same time as the arrival of 3,000 American reinforcements. Burgoyne, his men already in panic, had no option but to retreat in disarray back to their camp.

Fraser was taken back to his own lines, where he lingered in agony for more than twelve hours before dying at eight o'clock the following morning. In one of his final requests he sent word to Burgoyne that he wished to be buried on the battlefield at six that evening. Burgoyne acceded to his final wishes and he was duly buried in full view of the advancing American troops. As a mark of respect to a fallen warrior the American commander General Gates ordered that their guns be silenced.

There is some confusion over the order in which Timothy Murphy made his shots. Some say that he wounded Francis Clerke with his second shot before hitting Fraser with his third. Most sources suggest that Clerke was hit after Fraser. What no one seems to dispute is the distance involved: everyone agrees that the shot was made at over 300 yards in the heat of combat; a remarkable piece of marksmanship. It earned Murphy the nickname of 'Sure Shot Tim' among the other members of Morgan's Rifles.

The British defeat at Saratoga was one of the turning points of the Revolutionary War. His marksmanship in the second battle was only one of many exploits that made Timothy Murphy the stuff of frontier folklore. After his term with Morgan's Rifles ended he went back to his family farm and joined the local militia. He and a fellow officer were out scouting in Delaware Forest when they were captured by Indians. Although they were taken back to their camp and tied up, Murphy and his colleague managed to free themselves. They then methodically knifed all eleven members of the raiding party before returning to their own militia.

In many ways Timothy Murphy is a precursor of the modern sniper, a man who uses his deadly skill as a marksman to target opposing commanders and shatter the morale of the enemy.

Murphy was twice married and fathered a total of thirteen children. He eventually retired from the military and, although he never learned to read or write, he pursued a respected career in local politics.

He died of cancer in 1818 at the considerable age of sixty-seven.

A monument has been erected to Timothy Murphy on the site of the Second Battle of Saratoga where he made his famous shot. It was dedicated in 1929 by the then governor of New York State, Franklin Delano Roosevelt, who declared that the whole nation owed a debt to Timothy Murphy:

> This country has been made by Timothy Murphys, the men in the ranks. Conditions here called for the qualities of the heart and head that Tim Murphy had in abundance. Our histories should tell us more of the men in the ranks, for it was to them, more than to the generals, that we were indebted for our military victories.

3

THE SHOT NOT TAKEN

The potency of the sniper is that he is able to influence the course of a conflict with a single shot. But there are a few rare recorded occasions when the decision not to shoot has repercussions every bit as serious as pulling the trigger.

Major Patrick Ferguson was an Edinburgh-born British sharpshooter who had operated with some success against the Continental Army in the American Revolution. The British under General William Howe and their German allies had launched a campaign in 1777 to take Philadelphia and strike at the heart of the Americans. As the British made their approach, sailing along the Chesapeake, the American commander General George Washington believed they could be stopped at a place called Chadd's Ford, one of the few spots that would allow safe passage across the Brandywine Creek and on to Philadelphia.

Washington drew up his troops along the banks of the Brandywine in anticipation of confronting the British after they had landed at Chadd's Ford.

The British were equally keen to find out the

strength of the Continental forces, and on 7 September Patrick Ferguson and three sharpshooters were scouting the American lines. He and his men were operating under cover of darkness when they heard the sound of approaching hoofbeats. Taking cover they saw two horsemen coming towards them, one wearing a gaudy Hussar's uniform, the other the easily recognisable buff and blue of the Continental Army. Writing about the incident afterwards, Ferguson commented that the American officer, which is what he took him to be, was riding a bay horse and wearing a very large cocked hat. Ferguson also noted that the officer looked distinguished.

Certain that they could not be seen by the approaching officers, Ferguson admitted that his first thought was to kill them as they rode towards him. He then decided it would be better if he and his sharpshooters crept closer to them before firing. Stricken by shame at the cowardliness of his idea, Ferguson changed his mind again. He stepped out from where he had been hiding and stopped the two officers at gunpoint. The colourfully dressed officer was closest to him, but when Ferguson ordered him to stop he shouted a warning to his colleague. The American wheeled around on his great bay horse and galloped off in the direction of the American lines. Ferguson knew what he should do next but he could not bring himself to do it.

'As I was with the distance, at which in the quickest firing, I could have lodged a half a dozen balls in or about him before he was out of my reach, I had only to determine, but it was not pleasant to fire at the back of an unoffending individual who was acquitting himself coolly of his duty, and so I let him alone.'

The American officer galloped off towards safety without ever realising how Patrick Ferguson's honour had spared his life. A few days later, on 11 September, the British and the Continental armies clashed at the Battle of Brandywine. Patrick Ferguson's involvement in the action was short-lived, since he was hit in the right elbow by a musket ball.

Ferguson showed as much bravery in the hospital as he had on the battlefield. The musket ball had shattered his elbow and he had to undergo several operations, without anaesthetic, to remove bone splinters that had been driven into his arm. The pain must have been excruciating, but in a commendable display of noblesse oblige Ferguson made jokes about it in his letters home. It was in one of these letters, written in obvious discomfort and in a shaky left hand, that he revealed some remarkable news he had been given in hospital.

One of the surgeons who had been dressing the wounded rebel officers came in and told me that they had been informed that General Washington was all that day [7 September] with the light troops and only attended by a French officer in Hussar Dress, he himself, dressed and mounted in every point as described.

Ferguson believed that the man he had in his sights on the occasion of his patrol at dusk was none other than the American commander. There is no definite proof, but the circumstantial evidence suggests that Ferguson had good reason to think that he had been within seconds of killing the commander of the Continental Army, the man who would be the first president of independent America.

It does seem that Washington was out on a scouting mission that day, and he may well have crossed paths with Ferguson. The identity of the strikingly dressed Hussar remained a puzzle because there were no Hussars at Brandywine. However in March 1777 the Continental Army had begun recruiting foreign officers, of whom the most notable was probably the Scots naval commander John Paul Jones. These officers also included the Polish cavalry commander Count Casimir Pulaski, who had refused a commission because the Americans would not make him effectively second in command to Washington. Pulaski nevertheless volunteered and served as an aide de camp to Washington at Brandywine. He is, very likely, the man identified as the 'French Hussar' who was with the General that day.

Doubters suggest that surely Patrick Ferguson would have recognised George Washington by sight, but Washington in 1777 was not the icon that he is today; his face wasn't on any currency then. In addition Ferguson had only been in America since May 1777, so he may well not have been in a position to know what Washington looked like.

It seems likely that even if Ferguson had known the illustrious nature of his enemy he would still not have fired. He chided himself for his craven idea to shoot in the first place, and the idea of shooting a fleeing opponent in the back would have been anathema to someone of Ferguson's upbringing. He said as much himself in a letter home in January 1778 while still recuperating from his wounds. 'I am not sorry that I did not know all the time who it was,' said Ferguson.

The Battle of Brandywine Creek was a clear victory for the British. Washington appeared uncharacteristically naïve in his tactics and, the story goes, had to be warned by a local farmer that unless he moved his troops they would be surrounded. Washington had 11,000 troops under his command, and more than 1,000 were killed with a further 300 taken prisoner; he also lost eleven much-needed cannon. The British and Germans put around 12,500 men into the field, of whom around 600 were killed or wounded. The defeat at Brandywine forced the Continental Army to retreat and abandon Philadelphia to the British. Washington's troops drifted away, back to their homes and farms. The army dropped from a high of around 15,000 men before Brandywine to around 6,000 after the defeat.

Despite the defeat the Continental Army eventually rallied. Local communities provided what supplies they could and Congress sent reinforcements. None of this happened in time to save Philadelphia, but the Army still had its general and Washington was able to command his troops to victory. Whether, in the event Washington had perished at Brandywine, the war would have gone as well, or indeed whether America might still today be a member of the British Commonwealth, is a question for the counter-factual historians. But no matter what, America has good reason to be grateful to a Scottish infantry officer.

* * *

Patrick Ferguson's decency would have come as no surprise to those who knew him. He had been born in Edinburgh in 1744, the son of a prominent family that was at the heart of the Scottish Enlightenment. When he was twelve his father, a judge, bought him into the regiment of his uncle, Colonel James Murray, as an ensign. With war looming Patrick was deemed too young to serve, so he never took up the post. Later, when he was fifteen, his father bought him into the Scots Greys – the Royal North British Dragoons – as a cornet, the most junior rank available. The cost of the commission would have been around £800. Before he joined his regiment he spent two years studying at the Military Academy in Woolwich, where he became an expert on fortifications. After serving in the Seven Years War Ferguson fell ill with tuberculosis and had to return home. The disease would have the lasting effect of leaving him with bouts of near-crippling arthritis. By 1774 Ferguson was living in London. He had left the Scots Greys and bought a commission in the 70th Regiment of Foot, the regiment commanded by his cousin Lieutenant Colonel Alexander Johnstone.

As well as his lifelong interest in fortifications, Ferguson was also fascinated by the role of light infantry. At a training camp in Salisbury in 1774 he caught the attention of General Sir William Howe, who would later be his commander at Brandywine. Howe saw in Ferguson an aptitude for light infantry, and kept a close eye on his military progress in the United States. As a light infantry enthusiast himself, Howe would see the value of having Ferguson with him on his Chesapeake campaign. But before his commanding officer could make best use of him, Ferguson was wounded at Brandywine and any plans were stalled.

Despite his injuries, Ferguson was determined to return to combat. His right arm was now useless – the musket ball had left it permanently crooked – but during a painstaking six months he taught himself to ride, write, shoot, and even load with his good left hand. He declared himself ready to return to combat in May 1778, barely eight months after being shot. After distinguishing himself once again he was breveted to the rank of Lieutenant Colonel of the American Volunteers in 1780.

Linking up with Banastre Tarleton at the siege of Charlotte, Ferguson was highly effective in cutting off supply lines to the rebel troops. Later in the campaign he found himself appointed by General Cornwallis as head of the Loyalist Militia – that is those loyal to the king – in North Carolina marching to intercept a group of rebels, or patriots as they saw themselves, taking refuge in the dense Back Country. Ferguson and some 800 troops had set out from Gilbert Town on 27 September to confront the patriots. The initial rebel band had joined forces with other local Colonial militia preparing to wage a desperate war on Ferguson and his men, motivated by the frenzied rhetoric of local fundamentalist preachers as well as reports that Tarleton had refused to grant mercy to another rebel force and had massacred them.

Ferguson asked for reinforcements for his Loyalist Militia, but none could be spared. Instead he tried to rally as many men as possible that were still loyal to the king to join his forces. He was not above playing to the emotions of the crowd:

> . . . I say if you wish to be pinioned, robbed and murdered, and see your wives and daughters, in four days, abused by the dregs of mankind – in short if you wish or deserve to live and bear the name of men, grasp your arms in a moment and run to camp. The Back Water men have crossed the mountains . . . so that you know what you have to depend upon. If you choose to be pissed upon forever and ever by a set of mongrels, say so at once and let your women turn their backs upon you, and look out for real men to protect them.

Cornwallis advised retreat, and Ferguson knew that his only hope was in gathering enough men and women together and withdrawing south-east to the protection of Charlotte. It remained to be seen whether he could get there before the enemy struck. By 6 October he and his followers were about halfway back to Charlotte and camped for the night at King's Mountain. The following day they woke to find themselves face to face with a much stronger patriot force. Ferguson had 900 fighting men at his disposal; the patriot group comprised at least twice that figure. Outnumbered, Ferguson nonetheless spurned the chance to retreat, and battle was joined, with the patriots fighting so fiercely it was obvious they were seeking revenge for Tarleton's actions. Running low on ammunition, the Loyalists relied on bayonet charges to drive back the enemy. Three times they made successful charges, each time led by Ferguson, who was never far from the thick of the fighting as he drove his men on. Such were his efforts that two horses were shot out from under him in the heat of the battle.

Ferguson's forces had made their stand on a low ridge, and after an hour the patriots eventually broke through. They then drove towards his small command post at the other end of the ridge. Ferguson must have realised that the fight could not be won. He had seen his woman companion killed in the very first attack and knew there would be no mercy for him. He decided on one final attempt to break through the rebel lines. Swearing he would never yield to 'such damn'd banditti', he and two other volunteer officers, Colonel Vezey Husbands and Major Daniel Plummer, leaped into the fray once more. Husbands was killed at once, and Plummer was seriously wounded. Patrick Ferguson was blown out of his saddle as close to a dozen shots hit him. The rebel forces were delighted with their coup. His foot caught in the stirrups, Ferguson was dragged for some distance before his horse was finally stopped. Jubilant rebels blasted his corpse with shot after shot, stripped it and urinated on it, before allowing Ferguson's men to retrieve it and give it a decent burial. Many of Ferguson's men were hanged or tortured before the rebel rage subsided.

Banastre Tarleton had set out with reinforcements to try to reach his friend, but arrived at King's Mountain three days too late. It was through Tarleton that the rest of the army discovered Ferguson's fate:

... the death of the gallant Ferguson threw his whole corps into total confusion ... The mountaineers, it is reported, used every insult and indignity, after the action, towards the dead body of Major Ferguson, and exercised horrid cruelties on the prisoners that fell into their possession.

Patrick Ferguson was buried in a shallow grave on King's Mountain along with his lady friend, described as a buxom redhead and known as 'Virginia Sal'. Her real name may have been Sally Featherstone. A monument now stands on the site, but it bears his name only.

Ferguson's story contains one further twist. The news of the victory for the American forces was mentioned in dispatches on 27 October 1780:

The General has the pleasure to congratulate the army on an important advantage lately obtained in North Carolina over a corps of 1400 men, British troops and new Levies commanded by Colonel Ferguson. The militia of the neighboring country under Colonels Williams, Shelby, and others having assembled to the amount of 3000 men detached 1600 of their number on horseback to fall in with Ferguson's party on its march to Charlotte; they came up with them at a place called King's Mountain, advantageously posted, and gave them a total defeat; in which Colonel Ferguson with 150 of his men were killed, 800 made prisoners and 1500 stand of arms taken. On our part the loss was inconsiderable. We have only to regret that the brave Colonel Williams was mortally wounded.

The dispatch was signed by General George Washington, the man who could have died under Patrick Ferguson's hand three years earlier.

4

THE DAWN OF THE SHARPSHOOTER

Patrick Ferguson's contribution to the British Army was far greater than his deeds with the Loyalist militia. Although famous as the man who did not shoot Washington in the back, in his brief but distinguished career, more than anyone else in the British Army Patrick Ferguson is the man who made the work of the sniper possible.

At the beginning of the Seven Years War against the French in 1756 the British Army did not believe that the rifle, or rifled musket, had any place on the battlefield, at least not on the European battlefield. The British continued to put their faith in large bodies of highly drilled men exercising precision manoeuvres to engage the enemy in volley fire as they closed ranks. Indeed, a premium was put on not firing first: it was the epitome of courage and valour to take the enemy's first fusillades in your ranks before opening fire yourself. The British had few equals when it came to massed musketry. Rolling salvoes fired in company order along the length of the line could have a devastating effect on an enemy force. This tactic was so successful and so entrenched that in the mid-eighteenth century the Army's commanders did not see any great virtue in light infantry or skirmishers who could go out on patrol to scout and harass the enemy. The notion of individual soldiers being able to act of their own free will on the battlefield was completely alien and not a little threatening to the British military establishment.

There were, however, some free thinkers who realised that what could work in the colonies could work in Europe. These men could see the virtue of the rifle and the skirmisher. Among them was Patrick Ferguson.

Even in the late eighteenth century wars were still being fought in much the same way as they had been conducted at Megiddo, with two large, essentially static forces of men clashing in a prearranged location like great lumbering dinosaurs. These armies of the line were formed largely of traditional infantry who fought in regimented close-order formations three ranks deep. They were the 'heavy' troops, the men who marched into shot and shell with unquestioning determination.

However, a new type of soldier was beginning to appear on the battlefield. In 1740, the first year of the War of the Austrian Succession, the Austrians had called up troops from border outposts such as Croatia and Hungary. These men, not used to fighting in line formation, were 'irregulars', the contemporary successors of the slingers and bowmen of ancient times who had paved the way for the shock troops and their chariots. Able to fight and harry on the fringes of the battle, they would scout ahead and harass enemy forces to soften them up for the main force of the infantry. They became known as 'light' troops, semi-independent forces responding to calls from a bugle or a whistle to move with a great deal more autonomy than regular soldiers. The light troops often engaged the enemy first, providing a protective screen for the main army, a function that has been devolved to the twenty-first-century sniper.

It was the potential of light infantry that attracted Patrick Ferguson. The practice of employing light troops was not instantly adopted throughout Europe, but Ferguson believed that they could change the face of warfare. He was especially interested in light troops because they were the natural beneficiaries of another idea on which he had been working.

Ferguson, like every other serving soldier, was well aware of the shortcomings of the Brown Bess and realised that if British light infantry were to be successful then they would need a better weapon. He knew that the rifle would outperform the musket in firepower but would only be a genuinely popular battlefield weapon if it could be fired more easily and more quickly. He set about designing a breech-loading rifle for the British Army, taking as his model a design by the French gunmaker Isaac de la Chaumette. This design, developed in 1704, had been patented in England in 1721 when the Huguenot Chaumette fled across the English Channel.

Ferguson's conception was a rifle that would allow riflemen to load and fire faster than musket troops. During tests in 1776 he found it could be fired up to seven times in a minute, almost twice as fast as the smoothbore musket. Without any independent finance, Ferguson quickly ran up significant debts as he worked on his invention. He was forced to borrow from friends and relatives to raise the money to patent the weapon. Finally in December 1776 the patent was accepted for the Ferguson rifle. One of its early enthusiasts was the East India Company, which placed sufficient orders to help offset the large debts that had accumulated.

Before Ferguson's restless inventive mind turned to his next project – the development of a new field cannon – he was determined to convince the British military establishment that there was some merit in the rifle as a battlefield weapon. It probably helped that the Ferguson rifle was developed at the same time as the colonists were employing their rifles so effectively against British troops. Ferguson eventually found himself given permission to train an experimental unit of 200 men, but as British losses in the colonies continued to mount these orders were suddenly countermanded. Now he would have to make do with 100 men, and must leave for America as soon as possible.

Arriving in New York on 26 May 1777, the 33-year-old Ferguson realised that this was a rare opportunity. He had brought with him his experimental field gun, but this exploded on its first test because it was fired with the wrong type of ammunition. Instead, Ferguson concentrated all his efforts on his riflemen, who took to the field in what became a distinctive green uniform. This was Ferguson's suggestion to provide better camouflage than the screaming scarlet of the normal tunic, and was an adaptation of the more natural colours being worn by soldiers of European light companies.

The Ferguson rifle was one of the deadliest weapons in the British arsenal at the time. They should have been ordering more and reorganising their battlefield tactics around it. Instead, without the charismatic Ferguson, who returned to England to recover from his wounds on the Chesapeake Campaign, the rifle regiment was allowed to languish. It was an experiment that had been set up for one campaign only, but a fit Ferguson would surely have been able to lobby successfully for its continuation. The injury to Patrick Ferguson at Brandywine Creek was a serious blow to the development of rifle units in the British Army. Had his unit survived, given Ferguson's dash and pluck, they might have covered themselves in the further glory that their brief but impressive battle record suggested would follow them. Ferguson's untimely death three years later meant that the cause had lost a powerful advocate. Influential commanders such as General Howe were still supportive of the idea of light infantry, but without the exploits of adventurous soldiers such as Ferguson to prompt them, the idea remained an abstract notion and not a battlefield reality.

The British reluctance to use riflemen was not shared by other European armies. By the middle of the eighteenth century most had begun to include rifle regiments, which soon took an increasingly prominent place in the order of battle. In 1744, the first French rifle unit was set up, the Régiment des Arquebusiers. Later French rifle regiments would be known as 'chasseurs' – hunters, a name adopted for their German equivalent. Also in 1744 in Germany

Frederick the Great raised his Feldjäger-Corps zu Fuss – the Jägers. Frederick recruited his first unit from hunters and gamekeepers, the same groups who would provide a rich seam for sniper forces all over the world in later years.

The Jäger rifle was far from perfect. Although it was ideal for target shooting and hunting it was less well suited to combat. The barrel was too short for a bayonet for one thing, and it still took longer to load than a musket, which could, at times, leave the Jägers vulnerable, especially against large numbers of cavalry. Nonetheless it was still a vast improvement on the Brown Bess that the British Army had been using.

The British opposition to rifle regiments slowly began to weaken. Supporters of the idea could point out that previously in the colonies Britain had used locally recruited militia, armed with rifles, to good effect in the Seven Years War: among their number was one George Washington, who served with the Virginia militia. The continuing British experience some ten years later in the American War of Independence finally convinced them of the need for their own regiment of light troops. Eventually the Board of Ordnance took the leap of faith. In 1798 the British Army placed an order for 5,000 rifled muskets from Prussia. It appears that the batch that came to Britain was of very poor quality and did not perform as expected. There were enough however to equip the 5th Battalion of the 60th Regiment, which was basically an outfit formed from German Jäger troops who were fighting on the side of the British. This unit, in effect, became Britain's first rifle regiment.

The military establishment then took another brave step forward: it would test a variety of rifles to see which weapon best suited the needs of the British Army. A variety of weapons from all over the world were examined, along with the Jäger rifle, at Woolwich in 1799. These tests were extensive and exhaustive, and indicate that finally the issue was being taken seriously by the Army. The eventual winner was designed by a British gunsmith, Ezekiel Baker, who plied his trade in Whitechapel in the East End of London. He made weapons for the royal family and was under the patronage of the Prince of Wales.

Other weapons were more accurate, but the combination of an acceptable level of accuracy with a durable construction gave the Baker rifle the advantage over the others. Baker himself believed that the gun was at its most effective over 200 yards, although it could also hit a target at 300 yards if there was no wind. Contemporary tests showed that the Baker rifle could hit nine times out of twelve at a distance of 200 yards. Baker also claimed it could strike targets 400 or even 500 yards away, but not with any certainty of hitting what you were aiming at.

Unlike Patrick Ferguson's revolutionary breechloader, Baker's design was a traditional muzzleloading weapon but with a rifled barrel. It could be fitted with a bayonet, still the weapon of decision when the ammunition ran out or when the enemy closed in, which left its user less vulnerable on the battlefield. The Baker rifle also had the capacity to fire two different sizes of musket ball. The larger ball fitted more snugly into the grooves of the rifling, and was used when firing at distance to provide greater accuracy. But when the two sides had closed the distance between them, accuracy was less important than firepower. In that situation the Baker rifle could also fire a smaller-calibre musket ball, effectively converting it into a smoothbore infantry musket, which allowed a higher rate of fire. This duality gave the Baker rifle a rare tactical versatility. One of the more ingenious uses of the weapon by the British involved putting men armed with Baker rifles in the middle of the ranks of muskets. By using the smaller-calibre balls, the riflemen could pour on a significant rate of fire to cover for the muskets reloading. However if individual accuracy was required then it was a simple matter to switch to the larger ball and take more precise aim.

The decision was made after the 1799 trial to equip the British Army with the Baker rifle, and an order for 800 weapons was placed in 1802. Having taken the innovative step of ordering the Baker rifle the Army almost completely failed to take what might appear to be the next logical step: to adapt their tactics to the new weapon. Instead after adopting Baker's design for the entire Army, a single dedicated permanent rifle regiment was approved. Colonel Coote Manningham was authorised to set up the Experimental Corps of Riflemen in 1800, and this became the famous 95th Rifles when it was brought into line in 1802. But instead of using the new weapon to its maximum advantage and rearming and retraining the whole Army in its use, the rest of the British Army continued very much as before, as if the rifle were a musket, nothing more.

Like Patrick Ferguson before him, Manningham

selected men for their field craft and their hunting skills as well as their marksmanship. Unlike the rest of the Army, they did not wear the traditional red tunics. Manningham followed Ferguson's lead and had them outfitted in green. Throughout the nineteenth century, as rifle corps evolved in different armies across the world, so did traditions that were specific to riflemen. The green camouflage uniform was one such custom, as was the lack of regimental colours; these troops were conceived as skirmishers who were unlikely to need to advertise their presence in combat. Rifle regiments also marched at a much quicker pace than other units, to facilitate their rapid deployment, and since they were not normally operating as a single massed unit, they responded to whistles and bugle calls. The late Patrick Ferguson had been famous for his silver whistle, which he used to issue orders to his rifle corps in combat.

Slowly but surely the separate identities of the rifle regiments began to establish them as elite troops on the battlefield. This elitism seems to have been a perception by the regular soldiers; certainly the light troops received no concessions in terms of discipline or duties. They did however receive specific training, and were encouraged to act on their own initiative, which was alien to accepted military tactics of massed units of men moving around on command. If the Royal Navy's fast-moving frigates were 'the eyes of the fleet', then these skirmishers served the same function for the Army. Riflemen were encouraged to live off the land, to make use of natural cover, and to shoot sparingly but effectively. Operating in pairs or in teams of four they could range over the battlefield as they saw fit. Their only rule was that they did not surrender a position unless they could move to a better one.

Such skirmishers made deadly and dangerous enemies. They moved among enemy forces at will, staying out of range of their smoothbore weapons and picking off whoever they liked. Having confused and demoralised the enemy with their invisible firepower, the skirmishers could then use their field craft and training to simply melt away back into their own lines. They were snipers in everything but name, and they were among the first to prove the psychological effect of a sniper: the idea that a sniper might be in the area is enough to demoralise most troops and reduce their fighting effectiveness. It is no coincidence that the skirmishers, the forerunners of the conventional snipers, were the first to use the phrase 'One shot, one kill.' The grisly motto was a proud boast of the men of the 95th Rifles, who believed with some justification that their Baker rifles and their training made them more than a match for anyone they might meet.

Other European armies had their Jägers and their chasseurs, but none had a unit that functioned as radically as the 95th. Their comrades in other British regiments fought as part of the general mass of infantry that went where they were told and engaged the enemy on their officers' terms. The men of the 95th however had clearly defined aims and objectives on the battlefield. Their principal job was to forage ahead to find and kill enemy skirmishers. They would then turn their attention to artillery crews to silence enemy guns, and then to officers whose absence it was hoped would leave the enemy leaderless and chaotic.

Although the massed ranks that followed in their wake were grateful for the work of the skirmishers, a general dislike of snipers and sharpshooters by regular troops began to take hold. It began as resentment and would develop into suspicion, even hatred, over time. Opposing forces felt that sharpshooting from cover was ungentlemanly and dishonourable – and this dishonour transferred to the troops fighting with the benefit of sharpshooters' protection. Regular troops believed that the skirmishers additionally were being feather-bedded. It was, they felt, the conventional infantry regiments who took the risk of directly engaging the enemy, marching shoulder to shoulder in the face of shot and shell. The skirmishers were allowed to hide. The nascent antipathy would never really disappear.

5

'KISMET, HARDY . . .'

As the new century approached Britain found itself facing a fresh enemy. Napoleon Bonaparte had come to power in France in an effective coup d'état in 1799 that established him as one of three consuls who would govern the country. He quickly persuaded his fellow consuls to retire to private life, leaving him as First Consul and de facto ruler of the country. Napoleon's plans for the expansion of post-Revolutionary France inevitably included an invasion of Britain, but to achieve that he first needed to control the shipping lanes. He faced a considerable opponent: in the final years of the eighteenth century and the early years of the nineteenth, Horatio Nelson had become a torment to the French. Nelson's exploits in 1798 at Aboukir Bay, where he captured or destroyed all but two of Napoleon's ships, had made him a national hero. His reputation was reinforced by other victories, such as the Battle of Copenhagen in 1801, but his decisive battle came in 1805 when his pursuit of a joint French-Spanish fleet, which had taken him across the Atlantic and back, ended at Cape Trafalgar. Outmanoeuvred, outgunned and outmanned, the enemy fleet was crushed with appalling numbers of casualties. But at the moment of his greatest triumph Horatio Nelson was shot by a sniper on the deck of his flagship HMS *Victory*.

It was a common tactic for the French and Spanish to send marines armed with rifles up into the rigging when ships closed for battle. The British were less enthusiastic: Nelson in particular believed that the risk of a potential catastrophe caused by a stray shot hitting a barrel of gunpowder far outweighed any tactical advantage. The French had no such qualms. On the morning of 21 October 1805 HMS *Victory* engaged the French ship *Redoutable* off Cape Trafalgar and the two great warships yoked together with grappling hooks. As their cannon pounded across the 50-metre divide, the sharpshooters were sent aloft to perform exactly the same function as they performed on the battlefield: they picked off the officers and the gunnery crews. What happened next is not in doubt. How it happened remains something of a mystery.

The simple facts of the case are: Nelson was shot by a sniper perched on the mizzen mast of the *Redoutable*. The musket ball penetrated his left shoulder and went through his lung before lodging in his spine, crushing several vertebrae in the process.

'They have done for me at last Hardy,' Nelson told his flag officer, 'I am shot through.'

Nelson was carried from the deck of the *Victory* into the cockpit below, where he finally died of his wounds almost three hours later. The battle had been decisively won, the threat of a naval invasion had been emphatically ended, but Nelson was dead.

Who shot Horatio Nelson? In his memoirs *The Adventures of a French Sergeant*, published in 1826, Robert Guillemard, a French fusilier from Provence, claimed to have fired the fateful shot:

On the poop of the English vessel was an officer covered with orders and only one arm. From what I had heard of Nelson, I had no doubt that it was he . . . As I had received no orders to go down, and saw myself forgotten in the tops, I thought it my duty to fire on the poop of the English vessel, which I saw quite exposed and close to me. I could even have taken aim at the men I saw, but I fired at hazard among the groups of sailors and officers. All at once I saw great confusion on board the *Victory*; the men crowded round the officer whom I had taken for Nelson. He had just fallen, and was taken below, covered with a cloak. The agitation shown at this moment left me in no doubt that I had judged rightly, and that it really was the English Admiral. From the moment he received his wound, and the position of the wound itself, I could not doubt for a moment that I was the author; and I have ever since been fully convinced of it. But though the shot that had brought down this admiral had rendered a service to my country, I was far from considering it as an action of which I had a right to boast. Besides, in the general confusion, everyone could claim this honour.

Forensic historians have proved beyond doubt that the shot that killed the English commander did indeed come from the rigging of *Redoutable*; but there are claims that the French marines in the rigging were themselves shot only moments later by British sailors outraged by the shot fired at their Admiral. The famous painting *The Death of Nelson* by Maclise seems to bear this out. One sailor is clearly seen pointing towards the assassin while another British marine takes aim.

It is difficult to reconcile Guillemard's claims with Maclise's painting. Guillemard maintains he had no doubt that he fired at the famous Admiral Nelson. Could he really have been certain from the top of a swaying mast, shooting over a heaving throng of men engaged in close-quarters battle, through a shroud of dense smoke? Even if Guillemard was certain, then surely he would have come under return fire, as the painting depicts, and been unable to escape to tell the world his story.

The image of the avenging British sailors may simply have been propaganda aimed at placating the grieving British public. Daniel Maclise completed his work between 1859 and 1864, long after Guillemard had made his claim. The painter's purpose was to provoke sympathy for Nelson and all of the fallen at Trafalgar. However, Maclise is said to have researched his painting thoroughly. He interviewed many of the survivors of Trafalgar and based his work on their recollections.

Did Robert Guillemard really shoot Horatio Nelson? Was he even present in the rigging of *Redoutable*? Was the assassin immediately cut down himself? In many ways the death of Nelson parallels that of Lord Brooke at Lichfield almost two centuries previously.

The identity of the sniper remains uncertain. What is clear is that an influential and charismatic military figure was again lost to a single bullet.

Napoleon would never dominate the seas after his defeat at Trafalgar, but he continued his campaign on land, where he fought a new kind of war.

Before the nineteenth century war was still being waged as essentially a collision of large armies, with infantry, cavalry and cannon used as separate discrete units. Napoleon Bonaparte changed all that in 1799 by reorganising the French Army in his drive across Europe. Before Napoleon, battles such as that at Edgehill were often indecisive affairs, with the winner emerging after a long period of attrition. More attention was paid to outmanoeuvring an enemy or denying them advantage than actually inflicting defeat. Napoleon had no interest in that kind of warfare. He wanted his battles to be won, his victories to be total. He ordered fast marches interrupted by intense periods of fighting: tactics that were the forerunners of the German Blitzkrieg of the Second World War more than a century later. But to fight this kind of war Napoleon had to change the structure of the French Army. Adapting a system devised by Lazare Carnot, the Revolutionary Minister of War, Napoleon broke the Army down into divisions, each composed of units of cavalry, infantry, artillery, engineers and support troops. In effect each of these divisions became a mini-army in its own right.

His tactics depended on his skirmishers; fast-moving troops who could live off the land and operate, where necessary, in open order independent of their commanding officers. They were, literally, revolutionary. The new social structure of post-Revolutionary France made it possible for these principles to be accepted in the French Army. The British, still largely defined by class-bound social convention, thought of the common soldier as part of a barely disciplined rabble that existed solely to obey its social superiors in the officer class.

The only effective way to combat Napoleon's army was to fight like with like. British generals such as Sir John Moore were vocal in their demands that Britain recruit more light infantry and give them the training they needed. The military historian David Gates suggests that this sort of training was more psychological than practical. It wasn't so much that they needed to be taught how to fight; they had to be taught how to think for themselves. The French troops were perfectly capable of expressing themselves: they had fought a Revolution for the privilege. British troops however were part of an Army that was the last remnant of a feudal system in which everyone knew his place. Encouraging them to forget their social inferiority and to take the initiative by thinking for themselves was the key to unlocking their military potential.

Once this psychological hurdle had been overcome, British riflemen and skirmishers took to their new responsibilities with murderous vigour and soon proved themselves some of the deadliest men on the battlefield. The French soon came to despise the men of the 60th and the 95th, whose 'one shot, one kill' motto was proving to be graphically prophetic. One unnamed French officer gives a chilling account of what it was like to face them:

I was sent out to skirmish against some of those in green – grasshoppers, I call them – you call them Rifle Men. They were behind every bush and stone and soon made sad havoc amongst my men, killing all of the officers in my company and wounding myself without our being able to do them any injury.

During the Peninsular Campaign, Marshal Soult, the commander in chief of the French Army in Spain, wrote to the French Ministry of War on 1 September 1813. He complained bitterly about the King's Royal Rifle Corps, the old 60th:

... [they] are expressly ordered to pick off officers, especially field or general officers. Thus it has been observed that whenever a superior officer goes to the front during an action ... he is usually hit. A staff officer claimed that 'Les Riflemen' killed all our officers between July 25 and August 31, some 500 officers and 8 generals.

It was partly because of these devastating statistics that the French did not give quarter to any of 'Les Riflemen' they captured. Since the British shots were aimed, which the French considered to be unfair, they deserved to be judged as murderers. Soult was not alone in this belief: many British officers including the Duke of Wellington felt that sharpshooting or sniping was inherently indecent.

Nonetheless the British sharpshooters proved their worth in the Napoleonic Wars. The 60th and the 95th particularly distinguished themselves. There are many excellent eyewitness accounts of their activities, but few are better than the 1848 memoirs of Rifleman Benjamin Harris, a young Dorset shepherd who joined the 95th and was involved in some of the bitterest fighting of the Iberian Campaign of 1808–14. His first meeting with the French came at the Battle of Vimeiro in August 1808.

I remember our feeling towards the enemy that day was on the north side of friendly; for they had been firing on us Rifles very sharply, greatly outnumbering our skirmishers, and appearing inclined to drive us off the face of the earth . . . The grenadiers (the 70th I think) our men seemed to know well. They were all fine-looking young men wearing red shoulder-knots and tremendous looking moustaches. As they came swarming upon us, they rained a perfect shower of balls, which we returned quite as sharply. Whenever one of them was knocked over our men cried out 'There goes another of Boney's Invincibles.'

Harris paints a vivid picture of the reality of nineteenth-century warfare. It was a conflict of smoke and heat and pain and noise through which you had to defend yourself, hold your line and kill the enemy all without being able to see a hand in front of your face. It was of course the same for both sides, as Jean-Roche Coignet, a French version of Benjamin Harris, reported in his account of a battle at Allesandria in 1820:

... their numerous artillery overwhelmed us and we could hold out no longer. Our ranks were thinned visibly; all about us there were only wounded men to be seen, and the soldiers who bore them away did not return to their ranks; this weakened us very much. We had to yield ground. Their columns were constantly reinforced; no one came to our support. Our musket barrels were so hot that it became impossible to load for fear of igniting the cartridges. There was nothing for it but to piss into the barrels to cool them, and then to dry them by pouring in loose powder and setting it alight unrammed. Then as soon as we could fire again we retired in good order. Our cartridges were giving out and we had already lost an ambulance when the consular guard arrived with 800 men having their linen overalls filled with cartridges; they passed along our rear and gave us the cartridges. This saved our lives.

Although light troops were initially regarded as something of a luxury on the battlefield, the Napoleonic Wars confirmed their value. During the Battle of Rolica, Harris and his comrades in arms again found themselves embroiled in the thick of the fray.

I threw myself down behind a small bank where I lay so secure, that, although the Frenchmen's bullets fell pretty thickly around, I was enabled to knock several over without being dislodged; in fact I fired away every round I had in my pouch whilst lying on this spot ... The Rifles indeed fought well this day and we lost many men. They seemed in high spirits and delighted at having driven the enemy before them. Joseph Cochan was by my side loading and firing very industriously at this period of the day. Thirsting with heat and action, he lifted his canteen to his mouth; 'Here's to you old boy' he said as he took a pull at its contents. As he did so a bullet went through the canteen, and perforating his brain, killed him in a moment. Another man fell close to him almost immediately, struck by a ball in the thigh . . . I saw a man named Symmonds struck full in the face by a round shot, and he came to the ground a headless trunk. Meanwhile many large balls bounded along the ground amongst us so deliberately that we could occasionally evade them without difficulty.

From Harris's superb description of their fighting tactics it is obvious that the men of the 95th were becoming increasingly like the modern sniper. Harris talks about how he lay undetected and emptied a full pouch of cartridges into the French, killing them at will. His colleagues also adopted these stealth tactics. Elsewhere in his diaries Harris tells of one rifleman, a man called Jackman, who crept forward to the base of the walls of Flushing during the Walcheren Campaign in 1809. Jackman used a sword to dig himself an impromptu foxhole just outside the wall from which he was able to target the enemy.

. . . he laid himself and remained there alone, spite all of the efforts of the enemy and their various missiles to dislodge him. He was known, thus earthed, to have killed with the utmost coolness and deliberation, eleven of the French artillerymen, as they worked at their guns. As fast as they relieved each fallen comrade did Jackman pick them off; after which he took to his heels and got safe back to his comrades.

The rifleman came of age during the Peninsular War. One of the most vivid demonstrations of their value came during the British retreat towards the seaport of Corunna. The British, under Sir John Moore, were harassed and harried by the French as they struggled to make their way to safety. Eventually Moore made a stand at the Cacabelos Bridge just outside Villafranca. He left a small force of riflemen to guard the bridge and hold off the French while the rest of the Army was led into the shelter of the town. When Moore came back to the bridge he found his men coming under heavy fire from French dragoons led by General Baron Auguste Colbert. The British response was to take bets on who could shoot Colbert.

Eventually Thomas Plunkett of the 95th utilised the full potential of the Baker rifle. With a barrel measuring just 30 inches the Baker was shorter than a conventional weapon. Reloading Brown Bess would have meant standing up and revealing his position, but the Baker rifle could be reloaded and fired while lying down. Plunkett loaded his weapon and settled himself with his back to a small mound for support. Reclining until he was almost flat, he looped the sling of the rifle over his right foot in a textbook shooting position. From there, with his rifle braced for greater accuracy, he calmly took aim and shot Colbert cleanly through the head.

This sort of shooting was rare, but the Baker rifle and the tactics that it allowed to be employed changed the way the British Army fought its wars. Sniping or sharpshooting may have been ungentlemanly, but it had become a fact of war and a pragmatic necessity, whether it was artillerymen or French barons caught in the sights.

These new advances were not achieved without significant cost. The retreat to Corunna had left some regiments severely depleted. Rifleman Harris recalls that out of the nine hundred or so men who started the Peninsular Campaign only three hundred remained, most of them in a sick or weakened state. His own company had been one hundred strong but only three men, including him, survived. But as Harris phlegmatically observed, they recovered, their ranks were filled again, and they went on to fight another day.

6

A SHIFT IN POWER

Until 1849 the infantryman was the lowest form of human life on the battlefield. Before Napoleon injected rapid movement into his armies, warfare remained mired in ponderous manoeuvres to gain strategic advantage. Blocks of infantry moved around the battlefield like pawns on a chessboard. The balance of power lay firmly with the artillery and the cavalry; it was the artillery that would blow holes in the opposition ranks and the thundering cavalry charges that would break their spirit. All that was to change in 1849 when a French infantry officer, Captain Claude Minié, altered the face of warfare for ever.

Captain Minié's invention solved a long-standing problem by taking advantage of the spiral grooves – eight in the case of the Baker model – running along the length of a rifle barrel. These grooves allow the rifle to make better use of the laws of physics. Once a projectile is fired it essentially travels in an arc. The percussive energy of the powder charge is converted into the kinetic energy of the musket ball. From the moment it leaves the muzzle the ball loses energy, dropping at first imperceptibly then falling in a more and more pronounced manner until it eventually hits the ground after describing a shallow parabola. For maximum effect the target should lie somewhere along the arc of that parabola, preferably close to where the energy of the musket ball is at its greatest. When a smooth spherical projectile such as a conventional musket ball is fired, force is lost as the expanding gases of the charge escape around the ball. Additionally the soft lead ball itself had a tendency to deform as it travelled through the air, meaning that it lost power and accuracy in flight. It deformed further on impact. Sometimes musket balls ended up as no more than flattened lumps of metal that scarcely broke the skin of their intended targets: in such circumstances the musket was hardly a deadly weapon.

A spinning projectile will travel further and straighter than a smooth one thanks to its gyroscopic effect, and although it too travels in a parabola the energy lost is much more gradual. The Romans knew this: their more adept javelin throwers would wrap a cord round the end of their pilum so that this could be pulled as the spear was launched, imparting spin that made it fly straighter and truer. It was harder to make a musket ball spin: to make sure that it did, it had to be pushed into the barrel of the gun with considerable force, usually with a ramrod, to make sure it engaged the grooves on the bore. This took some time, even with innovations such as a greased patch round the musket ball to smooth its passage. In a target competition or a hunt, time was not critical; in the thick of a battle, anyone using a musket rifle was at a severe disadvantage. The conventional musket, such as the British Army's Brown Bess, could fire up to four times in a minute in the hands of a trained soldier, especially when loaded with a ball smaller than the bore of the musket. It could simply be dropped down the barrel, and what was lost in accuracy and power was compensated for by a greater rate of fire. A man using a rifle could be shot at several times by a conventional weapon while he fussed around trying to load his gun.

The rifle was a luxury on the battlefield, reserved for sharpshooters and skirmishers on special duty. The challenge lay in matching the accuracy of the rifle musket with the rate of fire of the smoothbore version.

The nineteenth century was an era of technical and engineering innovation in many fields. Not surprisingly it was also a halcyon age for weapon design. In the 1830s America led the way when firms such as Springfield and Colt introduced mass production to the art of gunmaking. Weapons that previously had to be hand-tooled and made by craftsmen were suddenly available in large quantities and relatively cheaply. Armies could be equipped with large numbers of long-barrelled rifle muskets. The arrival of mass production coincided with the development of the percussion cap; a charge that exploded when it was struck, for example by the hammer of a rifle. It did away with the frustration of trying to load and prime a weapon with loose black powder in all conditions.

As early as the sixteenth century, gunmakers had first wrestled successfully with the problem of finding a way to ignite the charge in the musket without using a match or fuse. The wheel-lock system was briefly in vogue for pistols; this was a mechanism whereby a revolving milled wheel struck iron pyrites suspended above the priming pan, which produced a spark to ignite the priming powder. The wheel-lock was a complex and delicate piece of equipment that was deemed too fragile and complicated to be entrusted to the ordinary infantryman. It was largely restricted to officers and cavalrymen, making it the first effective side-arm. Because it was not a robust piece of kit, it was eventually superseded by the snaphaunce mechanism devised by the Dutch. The word 'snaphaunce', from the Dutch 'snaphaan', means 'snapping cock', and indeed the mechanism did feature a much more sturdy spring-loaded hammer which levered downwards sharply, striking the flint to produce a spark. The snaphaunce also had a pan cover that would slide open before use, thus protecting the priming powder from the vagaries of climate. By doing away with the match, the snaphaunce eliminated the possibility of a soldier accidentally discharging his own or his comrade's musket. Troops could be grouped together more closely, increasing their combined firepower.

Further advances were introduced in the early seventeenth century by Gustavus Adolphus, king of Sweden, who revolutionised his own army with techniques that were quickly adopted by others. He reduced the weight of the musket, shortening the length of the barrel, and introduced the paper cartridge containing a fixed amount of powder with the musket ball attached to make sure the weapons performed consistently.

The snaphaunce mechanism gave way to the flint-lock by the middle of the seventeenth century; the generally accepted difference between the two is that the true flintlock is capable of being set at both half and full cock, and the cover of the priming pan is an integral part of the mechanism, sliding back automatically when the trigger is squeezed. During the English Civil War Cromwell's New Model Army had two companies armed with flintlocks, and the musket had been transformed into what we envisage today when we think of one. The flintlock was essentially a simple, reliable weapon that could also be fitted with a bayonet, so that armies no longer needed pikemen to protect soldiers as they reloaded or when their ammunition ran out.

One remaining drawback of the flintlock was the delay between pulling the trigger and firing the weapon, which disappeared with the arrival of the percussion cap.

Alexander Forsyth, a Scottish Presbyterian minister, was a keen hunter and amateur inventor who, like many other sports shooters, found that his hunting trips were frequently spoiled when his prey spotted the flash of the priming powder in the firing pan of his flintlock. They would frequently start to run in the few seconds it took for the priming powder to ignite the main charge and the musket ball to be fired. Forsyth experimented with a combination of fulminate of mercury and potassium chlorate in search of a priming charge that would ignite the main load so quickly as to be almost simultaneous. His idea was a small charge that would explode when it was struck and would set off the main charge almost immediately, thus eliminating the need for priming powder.

Forsyth's invention patented in 1807 was effective but still cumbersome, and remained of use only to sportsmen. However in 1814 Thomas Shaw of Philadelphia adapted Forsyth's ideas and came up with the percussion cap. Shaw's invention was essentially a copper cap – he had also experimented with iron and pewter – which was shaped like a hat with the priming charge in the 'crown' of the hat. When this was placed over the touchhole in the gun leading directly to the charge in the barrel, and then struck by the hammer of the firing mechanism, the gun would fire almost instantly. The percussion cap made life a lot easier for infantrymen, but it took more than twenty years for its military advantages to be appreciated. The British Army carried out extensive testing in 1834 before recommending its use. The 85th Light Infantry took percussion weapons to Canada during the Canadian Rebellion of 1837–8 and a Guards battalion took 800 percussion rifles that had been converted from flintlocks. By the end of the decade the percussion cap was also being used by France, Sweden and the United States.

The combination of a rifled barrel and an enclosed percussion cap went a long way to making the rifled musket a much more viable weapon. The rate of misfires decreased from just over 40 per cent to just over 1 per cent – two shots in every 200 on tests – and accuracy was improved from about 25 per cent of shots striking a target to just below 40 per cent. The rifle musket was now roughly ten times more accurate than the smoothbore version, and it could be made to fire more quickly. It was however still muzzle-loaded.

Once the rifle had been made faster and more accurate, attention now turned to the ammunition. How could the musket ball be made to spin while at the same time increasing the speed of loading and stopping the ball deforming in flight? The answer seemed to lie in making the ball engage more firmly and quickly with the grooves on the barrel. Bullet design consequently became an obsession with a group of early nineteenth-century inventors. The solution came in a radical redesign of the musket ball. Instead of the sphere that was prone to flatten, a new shape was used. The projectile was cylindro-conoidal, that is to say a cylinder with a cone on the top. The first such bullet was developed in 1823 by Captain Norton of the British 34th Infantry, who came up with a cylindro-conoidal bullet with a hollow base that would expand when fired to engage with the grooves of the barrel. This design was refined in 1841 by a gunsmith called Greener, who inserted a wedge, of either pewter or wood, into the cavity in the bottom of the bullet to provide greater stability.

Although it had shown a great deal of perspicacity and foresight with the testing of the percussion cap, the British military establishment quickly reverted to type and rejected Captain Norton's design as well as Greener's improvement. The French military were more farsighted, and they adopted Greener's design. Using this as a basis Claude Minié came up with a bullet that he patented in 1849.

This bullet quickly became known as the 'Minnie ball', although it was strictly speaking not a ball at all. The Minnie ball, like Norton's, is cylindro-conoidal: it looks like a modern bullet. It has three greased grooves at the top of the cylinder, and inside the base of the cylinder there is a cavity created by an iron cup. The bullet itself is made of lead. When the gun is fired the gases generated by the powder charge are channelled into the cavity, causing the soft lead of the bullet to expand. As they expand the grooves on the top of the cylinder are pushed into the grooves on the barrel of the rifle. This starts the bullet spinning as it travels along the barrel, imparting considerable force and velocity by the time it leaves the muzzle. Since the bullet is conical rather than spherical it is less likely to deform and will fly straighter and truer. The effects were remarkable. A Minnie ball could travel for up to 1,500 yards and was accurate at up to 350 yards.

There were a few initial problems. In some cases the iron cup inside the bullet would travel through the soft metal leaving the bullet in the gun. In other cases the iron cup had a tendency to drop out after about 50 yards and, since it was still travelling at the same velocity as the rest of the bullet, care had to be taken when firing over the heads of your own troops. Even allowing for these difficulties, Captain Minié had revolutionised the way that wars would be fought. His idea was not original – it was a variation on a design by Greener patented in 1841 using a pewter cup inside the bullet – but Minié's version was the best and quickly became the military standard.

The range and accuracy of the Minnie ball shifted the balance of power on the battlefield back to the infantryman. No longer was it necessary to wait until you were in a position where you were at risk yourself before you could fire your weapon. In the American War of Independence and the Napoleonic Wars musketmen could not fire effectively until they were about 100 yards from each other. The Minnie ball meant it was now easier to defend a position than attack: a unit armed with Minié rifles could fire at an attacking force more than three-quarters of a mile away and be certain of doing damage. With a Minié rifle, if you could see it you could hit it. The power of a Minié bullet was considerable. Tests at the time showed that at 600 yards a Minnie ball could penetrate six one-inch pine boards placed in front of each other. Suddenly the conventional frontal assault became suicide.

Cavalry also found themselves terribly exposed. The stirring cavalry charges of the Napoleonic Wars were now as wrong-headed as a frontal assault. Against a force armed with flintlocks or smoothbore muskets, cavalry could be used to devastating psychological effect to simply ride over the opposing troops. Faced with troops who could lay down rapid and deadly accurate fire from a quarter of a mile distant, cavalry became extremely vulnerable.

Artillery too suddenly found themselves imperilled. Napoleon, for example, was fond of putting his cannon in front of his troops to lay down support fire for his infantry soldiers, who would advance under the arc of his guns. Since many artillery pieces were smoothbore, Captain Minié's invention meant that riflemen had a potentially greater range than artillery: any artillery crew out in the open found themselves easy targets. The gun crews were forced to move to the rear, where they were less effective.

The effectiveness of the bloodcurdling bayonet charge also became more or less a thing of the past. An attacking force could only be sure of getting close enough to use its bayonets if it was prepared to sacrifice a large number of men crossing open ground through a hail of Minnie balls. Tragically, more than sixty years later during the First World War, when the Minié rifle had given way to the machine gun, there were generals who had still to learn the lesson.

After initially rejecting the design, the British Army was quick to issue the Minié bullet. By 1851, only two years after its patent, British forces had been re-equipped with Minié rifles. Although still not without some difficulty, loading and firing a Minié bullet was much easier than the complicated process of firing a conventional musket. The rifleman had a cartridge made of paper inside which was a bullet and a charge of black powder; usually about 60 grains, since the Minnie ball required more powder than a smoothbore weapon. The soldier simply bit off the end of the cartridge, poured the powder into the barrel and squeezed the bullet from its wrapping on top of the powder. This was then tamped down with a ramrod to make sure it fitted snugly and the rifle was ready to fire. Being a cone rather than a sphere made the bullet more aerodynamic, the perfect shape for travelling through the air. It was also less likely to flatten on impact, increasing its powers of penetration.

Britain was not alone in adopting the Minié rifle.

A general worldwide rearmament took place between about 1839 and 1850 as the world's great armies moved to percussion rifled muskets. The first recorded use of the Minié system in battle came with the British Army in 1852 during the Eighth Kaffir War. Combat reports from the period show that the Minié system was capable of dispersing the enemy at a range of 1,200 to 1,300 yards.

It was not just the range of the Minié system that made it so deadly; it was also its power. The consequences of a strike by a Minié bullet were horrific, as a Professor Longmore, an eminent surgeon, recounted in 1877.

If a modern rifle bullet, armed with its full force, strikes a hard and powerful long bone, like the femur for example, near the middle of its shaft it is broken into fragments of various shapes and dimensions often too numerous to be counted. A large proportion of these fragments are driven violently in various directions, and thus are converted into secondary missiles. A huge hollow is formed inside the limb which, when it is fully laid open and the effused blood washed away, offers to view a mass of lacerated muscle intimately mixed with sharp-pointed and jagged-edge splinters of bone. With all this extensive destruction within the limb, the external aspect of the wound through which the bullet first entered may exhibit nothing more to view than a small opening into which the top of the little finger enters with difficulty.

As well as the initial catastrophic damage, the internal cavities created in the human body by the path of the bullet were also inclined to quickly fill with blood and body fluids. If the wound was not treated quickly then sepsis set in, and the man was liable to die a slow, lingering death from infection as surely as if he had been killed instantly by the bullet.

The Minié bullet received widespread use during the Crimean War in 1853. Since the end of the Napoleonic Wars the British Army had been involved in almost thirty different wars, conflicts, rebellions and campaigns ranging from Nepal to Canada to New Zealand to Afghanistan. But the Crimea was the site of the first major European war since Waterloo when Britain joined its allies, Turkey and France, in attempting to limit Russia's expansionist ambitions in the Balkans and, by extension, Eastern Europe.

The Crimean War was as much a clash of technological levels as anything else. On one side were the largely pre-industrial Russian troops, on the other were combined British and French forces fairly bristling with the latest nineteenth-century developments in weapons technology. The effect of the Minié rifle was evident at the Alma. British troops had been trained to make good use of the Minié's range and accuracy. The Minnie balls tore through the massed ranks of the Russian troops: the power and range of the Minnie ball meant that a single shot could pass through five or six Russian soldiers. Later, at the siege of Sevastopol, men with Minié rifles were put in pits ahead of the British sappers in their siege lines ready to pick off any unwary Russian gunner who stuck a head above the parapet. This is, in effect, the first instance of what we now understand to be standard battlefield sniper tactics.

The effects of the Minnie ball were devastating. Anyone struck on the arm or leg by a Minié bullet could expect the bone to be broken or splintered for about ten inches either side of the impact, with additional damage done by the displaced bone fragments. Being shot in an arm or leg meant almost certain amputation if the victim did not bleed to death first. Anyone hit in the chest or abdomen was deemed to be beyond hope and left to die. Around 90 per cent of all the casualties in the American Civil War were caused by the Minié bullet.

Captain Minié's design nonetheless possessed some flaws: the rifle was prone to fouling because of a build-up of black powder. Although easier than it had been, loading the rifle still posed some difficulties, since the rifleman would either have to stand up or perform a complicated manoeuvre while lying prone. Also the rifle could not be fired undetected. The large charge of black powder required to fire a Minié bullet gave off a distinct puff of white smoke. Anyone shooting from a concealed position had to move quickly, his position no longer secret. It checked the effectiveness of the sniper.

One of the significant aspects of nineteenth-century warfare is how consistently technology responded to the demands created by its own innovation: it raced ahead of tactics. Commanders took a long time to realise that the methods of waging war had to change, and many thousands of lives were needlessly lost while lessons were slowly learned.

The loading problem was dealt with by the Prussian

inventor Johann Dreyse, who had developed several variants of breech-loading rifle between 1824 and 1836, and the Prussian Army took the bold step of issuing the weapon to its troops in 1841. It could not shoot as far as a Minié rifle, but what it lacked in range it made up for in rapidity, since it could be fired three times faster.

A smokeless propellant was also soon developed. From their earliest use in warfare, firearms had been distinguished by the production of clouds of thick, acrid, choking white smoke. Armies were often unable to see who or what they were fighting, as Rifleman Benjamin Harris recalls of his exploits at Vimeiro.

I myself was very soon so hotly engaged, loading and firing away, enveloped in the smoke I created, and the cloud which hung about me from the continued fire of my comrades, that I could see nothing for a few minutes but the red flash of my own piece amongst the white vapours clinging to my very clothes. This has often seemed to me the greatest drawback upon our present system of fighting; for whilst in such state, on a calm day, until some friendly breeze of wind clears the space around, a soldier knows no more of his position and what is about to happen in his front, or what *has* happened (even amongst his own companions) than the very dead lying around.

In 1886 French engineers replaced black powder with nitrated cellulose, and it proved an instant success. Unaffected by moisture, it burned quickly and fiercely, but, most importantly, it did not send up the telltale cloud of white smoke when the gun was fired. A wisp of grey smoke was the only evidence that a shot had been made. Alfred Nobel refined the process as well as developing smokeless explosives. This did much, quite literally, to eliminate the fog of war. No great clouds drifted across the field of fire: the age of the smokeless battlefield had arrived.

Captain Minié had made the rifle the most deadly weapon on the battlefield and the rifleman its most versatile combatant. With the introduction of smokeless propellants a rifleman could fire with accuracy from long range while remaining all but invisible. The conditions had been created for the modern sniper to enter the field of combat.

7

HIRAM BERDAN
AND HIS SHARPSHOOTERS

The American Civil War took place in a technological golden age of military development. The four years of conflict from 1861 to 1865 saw the introduction of magazine-loading rifles, machine guns, torpedoes, landmines, submarine mines, the telegraph, barbed wire, hand grenades, rockets, armoured trains, and the adaptation of the balloon to warfare. The Confederates used the mini-submarine and there were also plans in the pipeline on both sides for gas attacks, flame-throwers, and searchlights.

It was the first war of the industrial age, a conflict of machinery as much as it was of men, and in which the scientists and inventors on either side were almost as important as the troops. And it was one such inventor who provided the American Civil War's lasting contribution to battlefield tactics – sniper warfare.

Hiram Berdan was a colourful, some might say controversial, figure. He was born in Phelps, a small town in Ontario County in New York State, on 6 September 1824. Not much is known of what appears to have been an unremarkable early life except that he received a technical rather than a classical education. Berdan had a knack for machines; he seemed to have an intuitive understanding of how they worked. This gift for things mechanical and his skill as a mechanical engineer quickly made him a very wealthy man. With an active and inquiring mind coupled with the skill to turn his theories into practice, Berdan churned out a stream of military inventions. His developments included a revolutionary repeating rifle, a distance fuse for sharpnel, and a design for a torpedo boat. They added to his already considerable wealth. But Hiram Berdan did not become one of the most famous men in America because of his business acumen. Simply, Hiram Berdan was famous because he was the best shot in America. Above all other machines and technology, Berdan loved guns, and he was a natural crack shot. By popular acclaim he had been the best marksman in the country from 1845 to 1860 and he had the medals and trophies to prove it.

At the outbreak of the Civil War, Berdan began strenuously to lobby US President Abraham Lincoln to set up a regiment of crack shots to defend the Union. The leap from inventor to tactician was considerable, and it seems likely that the idea was not originally Berdan's own. He would doubtless have come across riflemen in competition who had experience of the Crimea or indeed the Napoleonic campaigns and were keen to swap stories. Or he could have gained the information via a more direct route.

Caspar Trepp, a Swiss marksman who had seen service in the Crimea as a skirmisher before emigrating to America, had conceived the notion of special units of marksmen with the best rifles who could advance the concept of the skirmisher. These uniquely trained men could function both as advance scouts or flank skirmishers and provide added protection for the remainder of the army. It could not have been easy for Trepp, an immigrant with no political connections, to overcome the inertia of the War Department. He attempted to lobby in a limited way by giving newspaper interviews in the summer of 1861, but eventually had to resort to offering his idea and his services to any man of influence who could back it. Hiram Berdan was such a man and quickly claimed it as his own.

Berdan's constant badgering of Abraham Lincoln revealed a serious character flaw. He may have been absolutely brilliant with machines, but he lacked subtlety when it came to dealing with men and women. In modern terms, Berdan just didn't have people skills. He seems to have been enormously egotistical. He was wealthy, bright, and a crack shot; he reasoned that the combination of these faculties ought to guarantee him power and influence. Each time he petitioned the authorities to set up a force of sharpshooters he asked for a suitably exalted rank for himself. Notwithstanding his lack of military experience or battlefield strategy, he expected to be put in charge.

In this respect Berdan seems to have been encouraged by General Winfield Scott, one of Lincoln's most senior commanders and military advisers. Possibly somewhat starstruck by Berdan and his achievements, Scott was more than happy to support his claims. On 14 June 1861, Scott's military secretary,

Lieutenant Colonel Hamilton, wrote to Berdan expressing the General's support:

> The General In Chief . . . desires me to say – that he was very favourably impressed with you personally – that a regiment of such sharp-shooters as are proposed by you and instructed according to your system, would be of great value and could be advantageously employed by him in the public service.

Once he had secured Scott's backing, it was almost certain that Berdan would get his way. Through his friendship with Scott, Berdan may have known he was pushing at an open door. The government in general and Abraham Lincoln in particular were sufficiently pragmatic to recognise the value of men who could shoot well enough to kill at a distance. There were already a number of sharpshooters among the presidential bodyguard, and on the day of Lincoln's inauguration the threat from the seditious South was deemed to be serious enough to station sharpshooters at key points on Capitol Hill.

Finally, on 14 June 1861 Berdan's special pleading paid off and he was given permission to raise a regiment of sharpshooters and to declare himself its colonel. The uncertain origins of the idea however would lead to a simmering resentment between Berdan and Caspar Trepp that periodically erupted into open rivalry.

When he came to choose the men who would wear the regiment's colours, Berdan wanted to be sure that he had the best and the brightest. The bar for entry was set deliberately high. During the summer of 1861 Berdan and his fellow officers travelled the breadth of the Union to visit events such as state fairs, town meetings and sports contests, looking for the best shots in the Union. Their arrival was announced in suitably strident terms in handbills and posters. Respecting the Jäger tradition, Berdan went out of his way to try to attract European marksmen. He made appeals through newspapers and speeches for Swiss and German hunters to join him.

Given the amount of publicity, there was never a shortage of applicants when Berdan's roadshow rolled into town. It was undoubtedly a glamorous unit, and its popularity was further enhanced by the belief that Berdan's troops were to be the best-paid in the Union Army. As well as a superior rate of pay they were to receive an additional weighting allowance from New York State. Berdan had also been told that his men would be equipped with the finest Sharps breech-loading repeater rifles available, a pledge that would soon become a sore point in a triangular argument between Berdan, his men, and the War Department. In the meantime many thousands of young men were seduced by the siren song of Berdan's recruiting pitch:

The government supplies each man with one of Berdan's improved Sharps rifles which will fire 1&¼ miles at the rate of 18 times a minute. We have no drill or picket duty. Our warfare is like the guerrilla or indian. Our uniform is green for summer, the color of grass & foliage, and Miller's Grey for fall and winter. You are privileged to lay upon the ground while shooting, picking your position. No commander while firing.

With the lure of freedom from discipline – and the suggestion that you need not be in harm's way – it's hardly surprising there were many more applicants than there were possible places. A shooting standard had to be devised. The test was as simple as it was challenging. All that the prospective entrant had to do was shoot at a target from a distance of 700 feet. The shooter could fire any rifle he chose from whatever position he preferred. If he could place ten shots into the target with none of them more than five inches away from the bull – in effect placing ten shots within a ten-inch circle – then he had passed the first stage of the entrance examination. The successful marksman then went on to take a second test, which was one of moral character and probity; it was not enough merely to shoot well. Berdan needed references and testimonials as to the good character of the men he was recruiting.

Even with these rigorous standards, by 24 September 1861 Colonel Berdan had 1,392 officers and men in ten full companies of what was duly constituted as the 1st United States Sharp Shooters. So high was the quality of applicants that only four days later Berdan was given permission to form another regiment. The 2nd United States Sharp Shooters contained 1,178 officers and men and was commanded by Colonel Henry Post. This second regiment would have been even larger but for the loyalty of the men of Massachusetts. The 2nd U.S.S.S. had been formed by companies from eight different states, but at the last moment two companies from Massachusetts decided to offer their guns to their own state, leaving the regiment with only eight companies instead of the usual ten. The Massachusetts men went their own way and became known as the 1st and 2nd companies of the Massachusetts Volunteer Sharpshooters. They served with distinction throughout the American Civil War with a number of other regiments.

In what seems to have been a concession to Berdan's rivalry with Caspar Trepp, the first company of the first regiment of United States Sharp Shooters, Company A, was made up almost entirely of Swiss living in the New York area, and Trepp was made its commanding officer. There were 136 men in Company A, and they would see some of the fiercest fighting in the Civil War before their tour of duty was over.

Allowing for the absence of the men from Massachusetts, Berdan had still managed to muster in three months a little more than 2,500 of the deadliest shots to be found in the United States. It was an achievement he was not slow in bringing to the attention of his Chief Executive in a letter to Lincoln on 21 July 1861:

The men are all selected, and all have come a long way inside of the requirements. I shall have ten companies ready in about two weeks. I have made the standard high – which has had the affect [sic] to bring out nearly all of the first class rifle shots. Should you call for three thousand 'sharp shooters' – all to come up to the requirements of those now being mustered in – I will

105

undertake to give them to you in thirty days and they would be worth thirty thousand common troops with the common weapons considering the mode of warfare adopted by the Rebels.

The first regiments of Sharpshooters were actually sworn in on 20 August 1861. Berdan had promised no less a person than the President they would be trained and equipped, so now he had to make good on that promise.

Hiram Berdan's personal vanity extended to the men under his command. They would not be wearing the traditional Union blue: Berdan was going to design a uniform for them. He came up with something suitably immodest, which added to the slightly Ruritanian air of the training camp.

Berdan's original plan was for his men to wear dark blue sack coats with metal buttons. The coat was intended to have a black fringe round the collar and hem, and be topped off with a soft felt hat with black feathers. He had a change of heart when he decided that the blue coats would make his men too conspicuous on the battlefield. Instead he adopted the idea of dressing his men in green. They wore forest green double-breasted frock coats with a lighter green piping. Their trousers were a blue-grey initially, but these were soon exchanged for a green that matched the coat. There was also a grey felt overcoat, which was soon discarded because it was cumbersome when wet and – more important – it was disturbingly similar to Confederate Grey. One of Berdan's men, a Lieutenant Colonel Ripley, voiced the concerns of the rest of the men: 'Certain grey overcoats and soft hats of the same rebellious hue were promptly exchanged for others of a colour in which they were less apt to be shot by mistake by their own friends.'

The ensemble was set off with a felt forage cap also in forest green. Many of the men adopted an added touch of a black ostrich plume set at a typically jaunty angle, but these accessories very quickly became casualties of war.

The contrast between Berdan's men in their dark green and the rest of the Union regiments in their workaday blue could not have been more pronounced. Berdan had wanted his men to stand out and he had got his wish. Their forest-green uniforms would allow them to blend in when they were going about the business of sniping, but on a parade ground with other regiments no one would be in any doubt which were the Sharpshooters. Not surprisingly, throughout their time in training camp they attracted huge amounts of attention from local newspapermen and foreign correspondents who recognised good copy when they saw it. The *New York Post* summed up the views of many when their reporter said that Berdan's men reminded him of Robin Hood and his band of outlaws.

The men of Berdan's regiments loved their splendidly unique uniforms. The regimental historian Captain C.A. Stevens spoke for all of them: 'By our dress we were known far and wide and the appellation of "green coats" was soon acquired.'

Berdan himself knew exactly what he was looking for in his men. He wanted a regiment that would be as immediately distinctive in battle as the colour of its uniform. Berdan's vision, although he had no military experience, was of a crack squad which would be capable of winning any fight for the Union:

> . . . to be effective Sharpshooters they had to be as skilled in field craft as they are in marksmanship, they must be self assured yet highly disciplined and above all they must be dedicated.

Although he set exacting standards and created a regiment that was admired and appreciated by Winfield Scott and Lincoln himself, that admiration does not seem to have been shared by the rest of the Union forces. The men of the other regiments of the Union Army seem to have looked on Berdan's men as something of a luxury. In a view that would be echoed for the next 150 years, there was a distinct feeling among the ordinary infantry that they would be doing the real soldiering while Berdan's men had the relatively easy job of sniping from the security of cover. The attitude of the other regiments is summed up in their nickname for Berdan's men. Both sharpshooter regiments wore cap badges with the initials U.S.S.S. picked out in brass inside a laurel wreath. The initials stood for United States Sharp Shooters, but as far as the other soldiers were concerned they stood for 'Unfortunate Soldiers Sadly Sold'.

The public at large, unaware of Berdan's machinations or self-promotion, took the glamorous new regiment to its heart. To them these men were not the U.S.S.S., they were 'Berdan's Sharpshooters'. There was even a popular song of the same name. It was sung to the tune of 'Yankee Doodle' and was no less patriotic.

Berdan's men are in the field:
The Rebels all look blue, sir;
For, well they know, where'er we go,
Our rifles will prove true, sir.

(*Chorus*)
Berdan's men are marching on,
Their deadly rifles rattling;
Swift, swift we dash through shot and shell,
For Freedom bravely battling!

O'er hills and through dark woods we roam,
No Foe shall make us tremble.
Our rifles crack, will drive them back,
And every fear dissemble.

Berdan's men etc . . .

As far as the citizens of the Northern States were concerned, Berdan's Sharpshooters were the answer to their prayers. These crack troops would cut a swathe through the forces of the upstart South and restore order to the country. There was no reason for anyone to think otherwise, unless they were actually in the Sharpshooters' ranks.

One of the best accounts of life in Berdan's Sharpshooters comes from Rudolf Aschmann, a young Swiss who volunteered for the regiment when he was barely twenty. Aschmann was the son of a judge and was born just outside Zurich. Like so many of his young countrymen he had been raised to hunt for sport almost all his life. He found the marksmanship test difficult but passed. Aschmann kept a diary during his time in the Civil War that was later published in Europe. He was assigned to Company A, and his support of his fellow countryman Caspar Trepp at times borders on hero worship. However his book is a generally pragmatic account of life with Berdan's men with little obvious attempt at self-aggrandisement.

It seems likely that Aschmann was induced to volunteer, along with so many others, by the glowing prospects that were held out for those who signed up. The recruiting posters and handbills painted a near-Utopian picture of life with Hiram Berdan. But life within the ranks of Berdan's Sharpshooters was nowhere near as glamorous as recruits might have been led to believe. They had been promised a fine gun, good wages, and an additional bounty from New York State. Very few of the commitments were met, according to Aschmann.

Except for the weapons, these were unfortunately nothing but promises, which gave rise to much discontent and bitterness later on. It may have been good in view of later events that many of us had been disillusioned right from the beginning, for we had to resign ourselves many a time during the war to seeing our hopes come to naught.

Company A was originally billeted in Weehawken, on the opposite shore of the Hudson River to New York. The men stayed with a local innkeeper, but space was very tight. When a second company of men who had been recruited in Albany moved in, conditions became extremely cramped. Racial ten-

110

sions between the Swiss and the Irish and Americans from the Albany Company frequently flared into quarrels and fights. Company A solved the problem in part by moving away from the inn and set up camp in a wooded area not far away. This camp became something of a European enclave, with its own tavern and makeshift streets named after Swiss heroes such as William Tell.

Eventually the 1st Regiment was reintegrated when all the companies were brought together in Washington for training and drilling. The men were to live in a village of streets made up of the tents of the various companies; there were ten rows of tents in all with about 35 feet between the rows. The field next to the camp was turned into a parade ground. Berdan and the other officers moved into a nearby house, which they converted to their headquarters and billet.

The men were generally well fed and well looked after in their new quarters, and the tensions and arguments that had previously dogged them seemed to dissipate once they had some room to move. Each soldier had a plentiful daily ration of beef, bacon and bread as well as generous allocations of coffee, beans, tea and sugar to be shared between them. They were worked hard during the day but their evenings were their own and life in the camp settled into a tough but bearable routine.

Life goes on even during a war, society continues to function and people still demand diversion and entertainment. In the autumn of 1861 and into the winter of 1862 one of the most popular entertainments available to polite Washington society was the training camp of the two new regiments of

Sharpshooters. Onlookers came from far and wide. There were society matrons, diplomats, visiting dignitaries, the mildly curious, even the President himself. They all had journeyed to see the cheapest and best show in town. Berdan drilled his men daily in front of large crowds of appreciative spectators. Target shooting was a popular sport at that time. It was how Berdan had become famous, and every day even larger crowds turned up for the regular displays of marksmanship and trick shooting. Little did they know that the ongoing row between Berdan and the War Department over weapons meant, at this stage at least, that the Sharpshooters were capable of little more than trick shooting. They had not enough guns to fight.

The weapons shortage confined the bulk of the two new regiments of Sharpshooters to their Washington training camp a good deal longer than they intended. Astonishingly, the War Department went along with the idea of setting up the regiments but balked at the cost of equipping them. The men had been promised the latest in weaponry, specifically the 1859 Sharps breech-loading rifle. This was loaded by pulling down a lever under the gun, inserting a cartridge, and pushing the lever back up again. It was a huge improvement on the old muzzle-loader. Although the 1859 Sharps had open sights it was still fearsomely accurate, and with a hair trigger it was a joy to shoot. The Sharpshooters had actually tested the gun during their stay at Weehawken and found it was far and away the best available to them. The problem was that the War Department had no intention of paying $45 a weapon for the Sharps. In real terms that was more than twice a sergeant's monthly salary of about $21 (privates received $13, and officers $105).

The intransigence of the Chief of Ordnance, General James W. Ripley, meant that the Union's latest crack regiments were effectively unarmed. The company from Michigan and the one from New Hampshire had the foresight to bring their own weapons. The others were left to make do with old muskets that provided them with just enough weapons to mount a picket as well as making sure that everyone was able to have some regular target practice. As long as the public could see them shoot on the parade ground then no one was any the wiser.

The War Department wanted to issue the men with five-shot revolving Colt rifles. These were good guns under normal circumstances, but there was a point of principle involved: whatever the qualities of the Colts they were not what Berdan's men had been promised when they signed up. Their argument was that if they accepted an inferior weapon then they would likely never receive what they had been promised. The debate over the choice of weapons raged through the winter of 1861/62. At one point the War Department tried to put the Sharpshooters into action at New Bern in North Carolina, but the officers and men refused to go without the proper guns. It was a three-way dispute. The soldiers complained to the officers, the officers complained to Berdan and Berdan complained to the War Department. Although he was very well connected in other areas, Berdan's political connections seem to have failed him.

Although the regiment was drawn from all across

the Northern States, with men of very diverse origins and backgrounds, the one thing that seems to have united the men is their attitude to Berdan himself. According to Aschmann he was not well liked by the rank and file, 'and often made a fool of himself with his clumsy demeanor. He had no military experience since he had never been associated with the military. He owed his position only to the influence of high-ranking personages and the favor of the President with whom he had been on good terms for a long time.'

The men were much more convinced by Lieutenant Colonel Mears, a regular army officer who had been assigned to the regiment. He knew that the men under his command were good shots but poor soldiers, and set to training them in the basic art of soldiering. He found a willing ally in Caspar Trepp, who may have seen this as an opportunity to embarrass Berdan.

One incident in the camp, recalled by Aschmann, highlights the attitudes of all sides. The company awoke one morning to find a caricature posted on the parade ground that caused much hilarity among the men. It showed Mears standing in front of a baker's oven with Trepp standing alongside him as his assistant. Mears was taking from the oven a tray on which lay Hiram Berdan in his full-dress regalia along with a few other officers who were perceived to be allies of his. The caption underneath said 'Freshly Baked Officers Available'.

The enlisted men loved the cartoon, especially since it turned out to have been drawn by the Sergeant of A Company. Berdan however was incandescent with fury, not just because he had been made to look a fool in the drawing but also because he was being told in no uncertain terms that the loyalty of his men really lay with Mears and Trepp. The furious Berdan launched an immediate investigation to discover the mystery artist, but even though everyone knew who had done it, no one turned him in.

Berdan's training programme was militarily weak, being more concerned with the Sharpshooter regiments' appearance on parade and exhibitions of target shooting in front of visiting dignitaries. Mears had been brought in specifically to address Berdan's shortcomings and fast-track the men to a point where they could be put into the line. Not long after they arrived in Washington the regiment suffered its first casualties when two men were wounded while on guard duty. Although their wounds were not fatal, the men had to be mustered out. Aschmann and others believed their comrades had been injured because they had not been well enough trained in the basics of soldiering.

Eventually under Mears the training began to take shape. They learned to take orders by bugle call. The men were also taught how to make use of the natural camouflage their green uniforms provided in woodland, and shown how to take cover during a battle. This was a novel concept: traditionally and to date, warfare still consisted of two lines of men advancing towards each other in plain sight before firing. As one of the Sharpshooters, William Greene of New Hampshire, put it, they were trained to 'run for the nearest tree and if there is none lay down on our bellies to get out of the rech [*sic*] of the enemie's [*sic*] fire'.

The monotony of training in Washington continued for the Sharpshooters while the war itself raged throughout the States. There was conflict in the War Department too, and not just about the merits of the Sharps against the Colt. Despite the evidence of the Crimea and the Napoleonic campaigns, and the technological developments that proceeded apace, there was still uncertainty about how the Sharpshooters should be used. Conditions may have been perfect for the sniper to enter the battlefield, but there were still those reluctant to throw him into the fray.

One suggestion, possibly supported by Berdan himself, was that the Sharpshooters should be used en masse. Their devastating long-range fire would decimate the Confederates as they charged across the fields towards Union lines. On the other hand one well-placed artillery barrage could wipe out a whole regiment of Sharpshooters. The influential Winfield Scott was among those who felt that they were far too valuable to use in the same place at the same time. His view was endorsed by the Secretary of War, Edwin Stanton. He believed the snipers should be organised into units no bigger than companies and attached to regular regiments to be assigned on a divisional basis as need dictated.

Stanton and Scott won the day. They also decided that the best shots in these two regiments of crack shots were to be detached for separate sniper duty. Their targets were Confederate officers, artillery crews, and other snipers. These men were issued with heavy 35 lb rifles with a special telescopic sight to make their deadly skills even more effective, and they had the absolute right to choose their targets. Finally an army had created a specialised sniping unit that we might recognise today.

In March of 1862 the order was finally given for Berdan's Sharpshooters to deploy. A compromise of sorts had been reached whereby they would be equipped at least initially with the Colt five-shot rifle. The 2nd Regiment was reportedly on the point of mutiny before they finally gave in and accepted the Colts. Once they had done this they were allocated to the Corps commanded by General McDowell. The 1st Regiment was less biddable and refused to accept the Colts. They were assigned to the Army of the Potomac under General George McLellan and left Washington on 20 March 1862 to carry out their new orders without any guns at all. They were finally placated and accepted the Colts, but only on the strict understanding that they be replaced as soon as the Sharps became available. Berdan had finally won his argument with the War Department, and the order to manufacture a consignment of Sharps rifles was signed in February 1862, almost six months after the regiments had been established.

The 1st Regiment United States Sharp Shooters saw their first action at the siege of Yorktown in the spring of 1862. They were thankful for the basic soldiering they had learned from Mears. The carnage of a Civil War battlefield bore no relation to the genteel rifle drills and turkey shoots back in Washington. The Confederates were using rifled cannon, which gave their artillery more distance and accuracy, and it was not uncommon for 80-pound shells to whiz completely over the Union lines, so powerful were the Southern guns. Berdan's men were put to work almost immediately trying to silence these guns. They were given carte blanche to go where they liked and do as they liked so long as the Confederate guns were silenced. Rudolf Aschmann recalled:

> There was more than enough work for our regiment . . . From small rifle pits dug in the dark of night as close as possible to the enemy batteries, we harassed his artillerymen with well-aimed bullets so much that they rarely showed themselves near their cannons anymore.

A popular song of the period, 'All Quiet Along the Potomac', published in *Harper's* magazine, stressed this new peril of death by sharpshooter:

> All quiet along the Potomac, they say
> Except now and then a stray picket
> Is shot as he walks on his beat to and fro
> By a rifleman hid in the thicket.
> 'Tis nothing, a private or two now and then
> Will not count in the news of the battle;
> Not an officer lost – only one of the men,
> Moaning out, all alone, the death rattle . . .

Being a Sharpshooter was dangerous work and the casualty rate was high, but there were compensations, many of which would become familiar to anyone carrying out the sniper's craft since then: 'we were exempt from all labour on earthworks and trenches, and we did not have night duty but were always relieved at nightfall when we could return to our camp.'

The value of Berdan's men to McLellan's forces was incalculable. After a very short time in the line they knew the position and strength of every Confederate gun. No sooner would one be fired than the crew would instantly expose itself to raking sniper fire. The news of the Sharpshooters' effectiveness spread and General William Smith, who was commanding the 2nd Division of McLellan's Corps on the left flank of the battle, requested their assistance. Confederate cannon were still very active on the left flank, so two companies of Sharpshooters, including Trepp's Company A with Rudolf Aschmann, were sent to assist.

They found themselves in a fight that would have been instantly recognisable fifty years later by any sniper on the Western Front. They were in water-filled trenches barely 200 yards from the enemy, firing through loopholes and afraid to move for fear of the slightest sound giving away their position to the enemy. Aschmann and his comrades found themselves having to go into the trench before dawn and leave after dark, which meant that there were days when they spent sixteen hours up to their knees in foul, stagnant water. As Aschmann noted, good results demand sacrifices.

> ... we carried out our tasks in a manner which brought us the praise even of our generals. The enemy marksmen could no longer continue unchecked and the artillery had to quit firing, for woe to him who showed himself even for an instant. His life was threatened by a dozen well-aimed bullets. The men who worked in the

rear of us could safely construct ramparts and entrenchments, and soon our enemies were surprised by full rounds of artillery fire.

As Aschmann describes it, another aspect of what would become textbook sniper tactics was evolving: the use of snipers to suppress and negate enemy artillery as well as to provide a tactical shield for your own troops.

Hiram Berdan was the sort of leader who took all of the credit and none of the blame, but the effectiveness of 'Berdan's Rifles' probably owed a great deal more to the training of Lieutenant Colonel Mears. The actual battlefield application of the troops seems to have been left to other more experienced soldiers such as McLellan and William Smith.

Even the Confederates recognised the value of Berdan's men, and in 1862 they passed an act of the Confederate Congress to set up their own sharpshooter regiment in response. The Confederate Sharpshooters were nowhere near as organised as Berdan's regiment, reflecting the general mood of the South about the imposition of external authority that had been one of the reasons for the secession. Like their colonial grandparents, many men of the South were good shots because they had to be if they were going to feed their families. They were also adept woodsmen and skilled in field craft. But men from this background were not inclined to allow themselves to be organised into regimented units. The best marksmen came from Mississippi, Tennessee and South Carolina, and they were selected by shooting competitions held throughout the Confederate Army.

The winners were named as sharpshooters and assigned individually to various units.

There was more than simply the status of the title of sharpshooter at stake in these competitions. While Berdan's men were itching to get their hands on the new Sharps, the Confederacy favoured the Whitworth rifle designed by the English gunmaker Joseph Whitworth. This design did not use a Minié-type bullet but instead derived its great power and accuracy from having a hexagonally rifled barrel and a similarly shaped bullet. This single-shot rifle was highly prized and, as well as being incredibly accurate, it was not as prone to fouling as the Minié system. The Whitworth had a range of at least 1,500 yards – one of its most famous shots was claimed at an even greater distance.

Since it was difficult to load and operate, the Whitworth was entrusted only to specialists. There was a second, more compelling reason for being sparing with the allocation of these rifles: the South simply did not have enough guns. In July 1861 the Union Navy had begun a blockade of the southern coast of the Confederacy, denying them the use of ports in Texas, Florida, Louisiana and South Carolina. This blockade would continue until the end of the war. Initially the Confederate Navy was ill equipped to deal with this threat, although eventually they would design and build smaller, faster ships that stood a chance of running the blockade.

The Whitworths had to be imported from England, and even with the nippier blockade-runners they were smuggled into the Confederacy with only limited success. The scarcity meant that a basic Whitworth could cost anywhere between $500 and $600, and one with a telescopic sight as much as $1,000. By comparison an Enfield rifle would cost around $150, and the Sharps $45.

Veterans recalled that the Whitworth shooters were treated as a precious resource that was not to be squandered.

> When firing these men were never in haste; the distance of a line of men, a horse, an artillery ammunition chest was carefully decided upon; the telescope adjusted along its arc to give the proper elevation; the gun rested against a tree, across a log, or in the fork of a rest-stick carried for that purpose. The terrible effect of such weapons, in the hands of the men who had been selected, one only from each infantry brigade because of his special merit as a soldier and skill as a marksman, can be imagined. They sent those bullets fatally 1200 yards.

The South never had more than 200 of these rifles at a time, but although their numbers were limited, in the right hands they were absolutely deadly and exacted a fearful price on Union forces. The Whitworth meant that effectively if you could see a mile, you could shoot a mile.

Blooded in combat at the siege of Yorktown, the 1st USSS were now moved on to Williamsburg, having first finally exchanged their Colt rifles for the new Sharps. Compared to the mechanically complex and hard-to-maintain Colts, the new guns, says

Aschmann, were everything they had been promised and more.

Besides being easy and quick to load in any position, they fired accurately even at great distances. They were easy to clean and keep in good working order, and more than any other gun in the army they had the look of a weapon worthy of a sharpshooter. They left nothing to be desired in soundness of quality, and soon this rifle came to be so well liked in the regiment that even the companies which were equipped with target rifles exchanged the latter for the new guns.

Nevertheless, despite the new guns, morale in the Sharpshooters gradually declined as the year went on. Since they were generally involved in the thick of the fighting, despite what other soldiers may have thought, they were beginning to take heavy casualties. At Fredericksburg in the winter of 1862, for example, the Confederates put Barksdale's Mississippians, one of their own sharpshooting regiments, into the line and inflicted serious losses on the Union forces in a withering sniping duel. By the end of the year the 2nd USSS was down to 110 men from an original establishment of more than 800. Some of this was caused by casualties but, since the bulk of the men were volunteers, there were also those who exercised their option of not re-enlisting at the end of their allotted tour. Henry Post, colonel of the 2nd Regiment, was among those who resigned.

The increasing tension in the 1st USSS Regiment between Berdan and Caspar Trepp was also having an effect on morale. Their long-simmering resentment erupted into the open when Berdan charged Trepp, who was by now a Lieutenant Colonel, with dereliction of duty. Trepp brought similar charges against Berdan and both faced inconclusive courts-martial. The official history of the regiment seems to implicitly criticise Berdan, albeit mildly:

> The enlisted men, they that handled the weapons that did the fighting, were silent lookers-on, wondering why their officers quarrelled so. Was this setting a proper example?

Against a background of their commanders' unflagging rivalry the Sharpshooters still went on to play an important role in the Battle of Gettysburg in July 1863. They were involved on all three days of the fighting and were responsible for the deaths of a great many Confederate soldiers. Although their role was nowhere near as great as Berdan claimed when he told people his men had turned the course of the battle, they were instrumental in the defence of Little Round Top against repeated Confederate attacks. They also served with distinction at Chancellorsville where, unusually, the 1st and 2nd USSS regiments fought together. Their combined efforts were enough to all but wipe out the 23rd Georgia Regiment and ensure they were mentioned in dispatches.

But Berdan could not repair the relationship with his officers. He came to believe that they, egged on by other generals who wanted to see him brought down, were complaining to the War Department behind his back. His paranoia was outlined in a long and rambling letter to the Secretary of War, Edwin M. Stanton. In a generally outraged tone, Berdan lamented the treatment he felt he was receiving from fellow officers and the War Department:

> . . . you will frankly inform me of all verbal statements that have been made to you; or from what you have observed, which in your judgment if true, would in any way reflect discredit on me as an officer and a gentleman.
>
> To say that I have made no enemies in the Army and especially in my command in two years active service, would in my judgment acknowledge that I have not done my duty as a commanding officer: but the time has arrived, when the interest of the service as well as my own reputation demand that I should do all in my power to remove any prejudice you may have against me, and with the hope that I may be successful at an early day.

Berdan's rift with Trepp was finally solved, tragically but inevitably, during the autumn campaign of 1863. The 1st USSS had taken heavy losses as it advanced along the Rapidan River towards Confederate forces. On 30 November they eventually came face to face with the enemy at Mine Run, a narrow but deep stream. The Union sharpshooters, acting as skirmishers, crossed the river to take control of Confederate rifle pits after driving out the Southerners. Trepp led the attack with Rudolf Aschmann at his side.

When Colonel Trepp and I were watching the enemy from one of these pits to give orders for further advance, an enemy bullet struck him down right at my side. The deadly lead entered his left temple, emerged above his right ear and rushed past, missing my head by only a few inches, then buried itself in the sand about ten steps from me.

Trepp did not die instantly. Although gravely wounded he remained unconscious but alive for several hours before finally succumbing. Caspar Trepp, the man who had originally come up with the idea of a regiment of sharpshooters, had been killed by a sniper. He was held in such regard among the Union forces that the divisional commander, General David Bell Birney, made a wagon available to bring his body back across the river so that he would not be buried on enemy territory. Trepp was eventually laid to rest in New York, with the Swiss community turning out in large numbers to honour a man they considered a hero.

Berdan himself was wounded a short time later and forced to leave the regiment to recuperate. The wound was serious enough to force him to retire from military service in January 1864. It also provided him with further reasons to plead his case to Lincoln, which he did in a letter of 21 February 1864 that tries to assert his place in history as the Sharpshooters' first commander:

You may remember that when you gave me this leave I remarked to you that I feared my wound would compel me to leave the service, & that you said in such an event you would give me a letter – that I had not had justice done me &c.

I had until recently hoped soon to be able to return to the service – but the Surgeons now inform me that this would be impossible that I could not pass the requisite examination. I therefore am compelled to give up all hopes of ever being able to render any more service to my Country in the field – I am still troubled with hemorrhage – caused by the motion of my wound in my lung when I cough or exhert myself in any way. I cant tell you how much I would value a letter from you to leave to my Children, showing that my services are appreciated by the Government I had fought so hard to sustain – It would be especially valuable to me from the fact that I doubtless lost this acknowledgment in the form of a promotion by sticking to my Sharp Shooters after the Secy of War & the Genrl in Chief decided not to Brigade Sharp Shooters – Also by offending the Secy of War by working for good Rifles, for my command, without which they would have been common Infantry, & with which they have rendered most important service –

You Mr President have always been my friend – & it is from you that I would be most proud to have a few lines, as you are not only fully aware of the troubles I have had in introducing this new arm of the service, but also as to what my commanding officers have said to you as to my conduct in the field.

Berdan hoped that Lincoln would recognise the great injustice he believed had been done to him and be keen to make restitution in the form of a glowing testimonial, which Berdan insisted was more for his children than himself. He also encouraged other friends and well-wishers to lobby the President on his behalf. No letter or communication from Lincoln survives, and since Berdan would almost certainly have made mention of it had there been such a letter we must assume that his request was ignored.

While Berdan was lobbying Lincoln again, what remained of the Sharpshooters were preparing for one of the bloodiest periods of fighting in the war. They fought in the Battle of the Wilderness and again immediately afterwards at Spotsylvania Courthouse as Union commander Ulysses Grant waged a relentlessly bloody campaign of attrition to break the fighting spirit of the Confederacy. Rudolf Aschmann recalls that of the 136 men who started in Company A, only twenty-six were left by the end of April 1864. They had barely another three months of their term of enlistment to serve, but in those three months they fought in eighteen engagements and lost another fourteen men.

The Sharpshooters killed more Confederates than any other regiment in the Union Army. They had proved beyond doubt that the sniper was a deadly and versatile component in the tactical arsenal of any army. The victory was not without a heavy price. Of the 2,570 Sharpshooters who enrolled during the war, 1,008 were either killed or wounded, giving them a casualty rate of around 40 per cent: appalling for a regiment that never charged the enemy or fought in closed ranks.

As Aschmann had suggested, by August 1864 most of the terms of enlistment for the men of the 1st USSS had run out and the regiment began to disband. The men who remained were reassigned to the 2nd USSS, which survived until 20 February 1865, when it too received orders to disband. Any soldiers still left were sent to units of their local state.

As the final units of Sharpshooters were broken up, they received a moving farewell from General Regis de Trobriand in General Order 12, issued on 16 February 1865:

The United States Sharpshooters, including the first and second consolidated battalions being about to be broken up as a distinct organisation in compliance with the order from the War Department, the Brigadier General commanding the division will not take leave of them without acknowledging their good and efficient service during about three years in the field. The United States sharpshooters leave behind them a glorious record in the Army of the Potomac since the first operation against Yorktown in 1862 up to Hatchers Run, and a few are the battles or engagements, where they did not make their mark. The Brigadier General commanding, who has had them under his command during most of the campaigns of 1863 and 1864, would be the last to forget their brave deeds during that period, and he feels assured that in the different organisations to which they may belong severally,

129

officers and men will show themselves worthy of their old reputation; with them the past will answer for the future.

It was a sonorous citation for history's first dedicated sniper force, one not matched by any specific citation for their vainglorious first commander. Hiram Berdan never received the exalted position he had been seeking. However, after the war he was breveted as a Brigadier General, United States Volunteers, for services at Chancellorsville and breveted as a Major General, United States Volunteers, for services rendered at the Battle of Gettysburg.

Berdan died on 31 March 1893 at the age of sixty-eight and is buried at Arlington National Cemetery.

8

'WHERE MAN DIES
FOR MAN . . .'

A martinet he may have been, and obsessed with his own advancement, but Hiram Berdan changed the face of warfare with every day the Sharpshooters spent in the field. The American Civil War was the first conflict to make wholesale use of snipers as a tactical weapon. For the first time they were deployed in large numbers with specific instructions to kill artillery crew, spotters, and especially opposing commanders.

One important distinction between Civil War snipers and their European counterparts was their attitude towards enemy officers. In the Napoleonic Wars and indeed in the Crimea, barely ten years before the American Civil War, there was a marked reluctance to shoot at officers. It was somehow seen as uncivilised: enlisted men and non-commissioned officers were fair game but ranking officers were gentlemen and not to be shot at. As they had done in the War of Independence, the Americans showed a highly practical disregard for such social niceties. Like the crossbow before them, the Sharps carbine and the Whitworth rifle were no respecters of rank and were impervious to social hierarchies. They were wholly democratic guns.

Commanders on both sides realised that there was often little innate tactical awareness in the volunteer soldiers and militias under their command. Apart from their raw courage, whatever effectiveness they had came from the craft and wisdom of the generals. If they could be removed, then the leaderless troops would find themselves in disarray. In consequence both Union and Confederate sharpshooters went out of their way to target officers. This tactic, radical at the time, became standard operating practice for the sniper in the Civil War and all wars that followed. It no longer mattered if you were outgunned, out-numbered, or out-thought. A sniper armed with a reliable rifle was a great equaliser.

There were few commanders in the war between the States as well loved by the troops under his com-mand as General John Sedgwick. Robert E. Lee was revered by his Southern troops, and although Sedgwick was not held in quite such high regard by his Union troops, many declared they would have followed him to the gates of Hell and beyond. Indeed in the brutal and bloody fighting at Antietam and the Wilderness some may have felt they had done just that. Sedgwick was known to his troops as 'Uncle John', a mark of the affection they felt for his strict but essentially fair-minded command.

Born in Cornwall Hollow, a small town in Con-necticut, on 13 September 1813, John Sedgwick was the son of a landowning farmer who was well respected in their small community. Sedgwick grew up with the benefits of a fuller than usual education:

he spent some time in secondary school. Most of his further education was gained at the hands of the Reverend William Andrews, a colleague of Sedgwick's father Benjamin, who was a deacon. From his father John Sedgwick inherited a strict sense of discipline and fair play; he also inherited an impressive physique that in later life gave him a somewhat bear-like appearance.

The young Sedgwick taught in the local school himself for two years before he was accepted for West Point. No better than an average student, he did not distinguish himself at the military academy and graduated in the middle of his class in 1837. Sedgwick's talents lay not in the dry lectures and dusty theorising of the classroom but in the heat of the battlefield. It was here, in the chaos of shot and shell, that his cool and analytical manner brought him distinction.

Sedgwick began his career fighting Indians. He was immediately commissioned as a second lieutenant of artillery, where he fought first in the Seminole War and then the following year in the inglorious campaign against the Cherokee. Settlers had been arriving in Georgia in droves, taking more and more of the land owned by the Cherokee. Then with the discovery of gold in North Georgia the US Army became involved in what was effectively a mass resettlement of the Cherokee nation from Georgia to Oklahoma, a distance of more than 1,000 miles.

After a brief posting to America's border with Canada in 1846, Sedgwick began to make a name for himself in the Mexican campaign of 1846–8 at Veracruz, Cerro Gordo, Puebla, Cherubusco and

Molino Del Rey. He was breveted as a Captain at Cherubusco and later as a Major at Chapultepec. His commission as a Captain was finally confirmed in 1848, when he was made captain of Duncan's Battery of Light Artillery. His commanding officer, Colonel Duncan, was heard to remark that Sedgwick was the coolest officer under fire he had ever seen. There were those who predicted that he would rise to become Commander of the Army.

Sedgwick spent the next few years fighting Indians on the Western frontier, but was recalled to the East at the outbreak of the Civil War. During the constitutional crisis when the South seceded he was twice promoted to replace Robert E. Lee, then a Union officer. The first time was when Lee himself was promoted and the second came when Lee resigned his commission because he could not raise his sword against his beloved home state of Virginia. A Democrat by inclination, Sedgwick chose to fight on the side of the Union. In a rare visit back to Connecticut – Sedgwick took only three furloughs in almost thirty years of service – his father remarked that the Civil War might be a good opportunity for Sedgwick to further his career. In an uncharacteristic moment of boastfulness the younger man allowed that he might make full colonel by the time it ended.

In fact he was a colonel almost by the time it started. Although he was promoted very quickly, his opportunity to distinguish himself in combat did not come quite so quickly. Illness confined him to quarters and he completely missed the first Battle of Bull Run in 1861. By the following year, much recovered, he was a brigadier with some 13,000 men under his command. In a crucial encounter at Fair Oaks in the Shenandoah Valley in the summer of 1862, Sedgwick was credited with saving the Army by a famous river crossing under near impossible conditions.

After his heroism at Fair Oaks, Sedgwick was promoted to major general and led his 9th Corps into battle at Antietam in September 1862. The engagement at Antietam was the bloodiest encounter on home soil in US history. By the end of the day on 17 September both sides had lost a total of almost 5,000 men with a further 20,000 wounded and around 3,000 missing or unaccounted for. Unwittingly Sedgwick marched his men into a Confederate trap. Attacked from three sides, he lost half his men, had his horse shot out from under him twice, and was wounded three times. He had to be carried from the field barely alive.

Antietam left scars on all who fought there, not just those physical ones that sent Sedgwick back to Cornwall Hollow to recuperate, although he returned to command the 9th Corps before he was fully recovered. He was once heard to remark in a darker moment: 'If I am ever hit again, I hope it will settle me at once. I want no more wounds.' It was a remarkably prescient wish. In the meantime, so delighted were his fellow officers to see their commander returned to them that they presented him with a dress sword that they had especially made in France at a cost of $1,000, a staggering sum in those days when a sergeant earned a little over $21 a month.

Their affection notwithstanding, Sedgwick was soon transferred and given command of the famous

6th Corps. Again he distinguished himself. At Fredericksburg, when the Union Army was being held by Confederate sharpshooters and snipers, it was Sedgwick who led his own 6th Corps, along with the 1st and 3rd, in a crossing of the Rappahannock River that came to be regarded as a textbook military manoeuvre. Facing most of Robert E. Lee's Army and without the support he expected, he executed a magnificent fighting retreat that saved the bulk of his men. Later Sedgwick was one of the few Union commanders to emerge with any credit from the disastrous Chancellorsville Campaign which, combined with his distinguished service, doubtless led to him retaining his position when the five corps of the Union Army were reduced to three.

There is no doubt that Sedgwick was a brilliant officer. His forced march to Gettysburg was one remarkable achievement among many. It is ironic then that he is chiefly remembered not so much for the accomplishments of his life but the black comedy of his death.

In May 1864 a small town barely 60 miles from Washington, Spotsylvania Courthouse, had assumed major tactical importance in Ulysses Grant's campaign of gruesome attrition against Robert E. Lee. It was Lee with his smaller force who won the race to Spotsylvania, arriving there on 8 May. Sedgwick's 6th Corps made a rapid march and arrived just outside the town at around five in the evening of the same day. The 6th Corps was supposed to support General Gouverneur Warren's men and Sedgwick spent the rest of the evening getting his troops settled in before riding back to Warren's headquarters. Along with the rest of the command, Sedgwick spent the night under the stars.

The following morning just after daybreak Sedgwick began to set out his line of battle, busying himself with various tedious but necessary chores while his men dug in. The best account of what followed comes from Sedgwick's loyal chief of staff, Martin T. McMahon who was a brevet major general with 6th Corps. Sedgwick was noted for his sense of humour and McMahon recalls how he settled himself on a couple of provisions boxes and started to tease his junior officers. McMahon believed that Sedgwick was building up to playing some trick on the other men when he suddenly stopped in mid-flow. His tactician's mind had noticed that the rifle pits into which men were filing in front of him were at the wrong angle and would be exposed to a Confederate artillery battery.

Moving the men was a job for McMahon, not for Sedgwick. None the less, when McMahon got up Sedgwick went with him. The two men mounted their horses and rode casually towards the offending troop emplacement. McMahon recalls that the previous day he had half-jokingly warned his commanding officer not to go near the artillery pieces, pointing out that every officer who had shown himself near them had been shot. At the time Sedgwick had taken the advice and, to console McMahon, said he could think of no reason to even go near the battery. Both officers had forgotten this conversation as they trotted towards the men in the rifle pits.

As McMahon and Sedgwick showed themselves there was, as McMahon had predicted, a smattering of sniper fire from the Confederate lines. The Union troops ducked and dodged as they changed position. McMahon later recalled Sedgwick's famous coolness under fire coming to the fore:

> The general said laughingly 'What! What! Men, dodging this way for single bullets! What will you do when they open fire along the whole line? I am ashamed of you. They couldn't hit an elephant at this distance.'

Sedgwick's chiding had little effect on his troops, though some were doubtless chastened. Moments later a soldier separated from his unit passed right in front of Sedgwick at the same time as they heard the distinctive whistle of a Confederate bullet. The man dropped to the ground for protection and McMahon takes up the story:

> The general touched him gently with his foot and said 'Why, my man, I am ashamed of you dodging that way,' and repeated his previous remark, 'They couldn't hit an elephant at this distance.' The man rose and saluted and said good-naturedly, 'General, I dodged a shell once, and if I hadn't it would have taken my head off. I believe in dodging.' The general laughed and replied, 'All right, my man; go to your place.'

McMahon recalls that once again he heard the long shrill scream of a bullet. This time it did not whirr harmlessly past, its journey was cut short by the distinctive sickening slap of metal bursting through flesh. It interrupted his conversation with his commander.

. . . when, as I was about to resume, the general's face turned slowly to me, the blood spurting from his left cheek under the eye in a steady stream. He fell in my direction; I was so close to him that my effort to support him failed, and I fell with him.

McMahon's shocked cries were heard by a group of officers standing nearby, including a surgeon. These men rushed to their fallen general.

A smile remained on his lips but he did not speak. The doctor poured water from a canteen over the general's face. The blood still poured upward in a little fountain.

It was a mortal wound. Sedgwick, who had asked that he should have no more wounds but be settled at once, had got his wish. He died on the field of honour without uttering a sound. It is to his credit that his men, though they knew of his plight, did not break ranks. They knelt at attention, maintaining the discipline he had instilled in them, craning their heads to hear of any news of their beloved general. The anguish of the troops at the death of their commanding officer is eloquently portrayed in a painting by Julian Scott that hangs in Plainfield, New Jersey. In a scene that evokes the Pietà, anxious enlisted men and officers look on in grave concern as the desperately wounded general breathes his last in front of them.

More details of Sedgwick's death emerged from the Confederate side in an account by a soldier of the South Carolina Sharpshooters:

This very same day Ben Powell came in and told us that he had killed, or wounded, a high-ranking Yankee officer. He said that he had fired at a very long range at a group of horsemen that he recognised as officers. At his shot one fell from his horse and the others dismounted and bore him away. That night the enemy's pickets called over to ours that General John Sedgwick, commanding the 6th Corps was killed that day by a Confederate sharpshooter.

It seems likely that Sedgwick was the officer killed by Powell, a Confederate sniper. Some sources put the distance of the shot at up to a mile, but from McMahon's eyewitness account it seems the gunman was firing from around 1,000 yards. He would certainly have been using a Whitworth rifle. It was a fine shot and had a devastating effect on the Union forces. Sedgwick had been arguably their best and brightest and was almost irreplaceable. Shattered by the death of a much-loved officer they were unable to press home any advantage they might have gained. After thirteen days of bitter fighting the two sides left Spotsylvania without either gaining a decisive victory.

Sedgwick's corpse was carried back to the Union lines and General Meade's headquarters. It was not, as was the custom, taken into whatever house had been pressed into service as a makeshift hospital-cum-mortuary. Instead his men cut down trees and shrubs to build a bower where his body lay on a bier of leaves and branches. Some days later he was taken to New York, where his body lay in state under a wreath provided by Mrs Lincoln, the President's wife. On

Sunday 15 May 1864 he was buried in the cemetery at Cornwall Hollow, the town of his birth, where a monument now stands in his memory. Listing his battle honours it describes him as a skilled soldier, a brave leader, a beloved commander, and a loyal gentleman. It bears a simple inscription: 'The fittest place where man can die is where man dies for man.'

Beyond the specialist ranks of Civil War historians, John Sedgwick is remembered as a footnote in trivia books. His misquoted last words are his lasting echo. It is more convenient to promote his chiding about the enemy's inability to hit an elephant to his final utterance, rather than the more banal words addressed to a soldier who showed a caution the General would have been wiser to follow.

The Confederate snipers did not know they were shooting at Sedgwick specifically: they were unaware he was dead until they were informed by the Union sentries. That spot, as Martin McMahon remembers, had proved particularly fruitful for them in terms of Union officers struck down over the previous days. They were simply aiming to take out commanders and scored a spectacular success. Had Sedgwick remembered his promise to McMahon not to put himself in harm's way, the stalled Union forces, with Sedgwick to rally and lead them, might have been victorious at Spotsylvania. Had they won there, the war would surely have been shortened.

John Sedgwick was probably the highest-profile casualty of the American Civil War to fall victim to an opposing sharpshooter, but he was by no means the only one. General John Reynolds was killed by a sharpshooter at Gettysburg and General William

Lyttle was killed as he led a charge at Chickamauga by a sniper with a Whitworth. At one stage in the war the Confederate Whitworth men were killing or wounding so many Union officers that orders were issued to reduce the size of their badges so they could not be so easily identified. This anticipated the way wars would be fought in the future. Thanks to the sniper, generals and other high-ranking officers retreated to the rear; no longer would they command from the front. The day would soon come when they would not venture from their command post at all. No longer would the generals die gloriously in battle. In future they would die in their beds.

The loss of a brilliant commander could cause disarray in the field through the action of a patient sniper dedicated to his task. Hence, almost as prized as an enemy officer was an enemy sniper. The Civil War not only saw the introduction of organised sniping, it also saw the beginning of counter-sniping operations.

One well-documented sharpshooter duel took place in June 1864 at Kennesaw, where Union snipers were wreaking havoc among the ranks of the Confederate Orphan Brigade, especially among their skirmishers. Eventually a veteran soldier, Virginius Hutchen, took matters into his own hands. He made his way forward, passing several former comrades from the Kentucky regiment each lying dead with a single gunshot wound to the head. He asked where the men had been fighting and was directed towards a pile or rocks beside a tree. After some thought Hutchen took up a position a little way from the makeshift fortification.

Orphan Brigade historian William C. Davis retells the duel that followed:

> Shortly he crawled to the rocks, put his hat atop a stick, and held it above the breastwork. Instantly a bullet perforated his headgear. Three more times he repeated this charade until he spotted a tuft of smoke that betrayed the enemy marksman's position. At that moment Taylor McCoy of the brigade sharpshooters approached, and Hutchen borrowed his Kerr rifle. One shot brought down the enemy picket and Hutchen felt well satisfied in having, however slightly, evened the score for his departed comrades.

It was during such sniper duels that the tactics that would become standard counter-sniper operations first evolved. A forage cap would be raised on a stick to draw fire, in the hope of inducing a sniper to reveal his position; snipers would lie in camouflaged hides for hours at a time waiting for the right shot; they would also fire from roughly fortified positions in which they created crude loopholes to allow the shot to be taken. For these men war was no longer an impersonal contest of charging, heaving masses. These sniper 'duels' were well named. Thanks to the Whitworth and the Sharps, war for the sharpshooter had become as personal and intimate as the settling of a grievance at dawn with pistols and seconds.

If snipers on either side seemed to be without any qualms about the manner in which they took lives, they did at least have a respect and regard for their opposite numbers. As the war ground on, troops in the field would occasionally impose their own ad hoc truce. Rudolf Aschmann recalls one incident during the siege of Petersburg when a lull in the fighting was rudely broken:

One day when we were on picket duty a shot fell from the Confederate side without, however, striking anyone on our side. Indignant at this breach of the peace we made ready to shoot back but the enemy immediately gave a sign that this disturbance was due to a mistake. Soon after, the delinquent appeared in front of the enemy line where as a penalty he had to parade back and forth for two hours in plain view of both picket teams, carrying a heavy beam on his shoulder. Loud shouts of applause indicated to the Southerners that this was satisfaction enough and that peace had been completely restored.

There were times, though, when the troops had no option but to fire. Even then, as Aschmann recalls, there was plenty of evidence of cooperation between the opposing forces:

Occasionally the higher command of one or the other side would give orders to open fire on the opposite patrols. In such cases they usually gave each other a warning signal, and once we even heard a Southerner call 'Yankees, lie down! We have orders to shoot!' Immediately fire was opened and kept up all day with intensity.

If the sharpshooters treated each other with professional courtesy, they did not expect to be treated in the same way by other troops. A sharpshooter who was captured by the enemy expected to be summarily hanged, another tradition that would survive through the years. At one stage Union troops would hang anyone they found with a Whitworth rifle. Imagine the relief then of some Whitworth men at Gettysburg who were captured and were expecting the death sentence when they discovered they had been captured by Berdan's men.

Feelings ran high on both sides. During the siege of Yorktown one of Berdan's Sharpshooters, Private John Ide, found himself in a deadly long-range sniper duel with one of his Confederate counterparts. Ide was using a target rifle with a telescopic sight and firing from the side of an outhouse. The duel had been going on for so long that other Union troops stopped what they were doing to watch. They saw Ide, about to raise his rifle once more, felled by a shot from the Confederate sniper. It was a killing shot, right in the middle of his forehead. It was met by a huge cheer from the Confederate lines, who were also observing this single combat of the snipers like spectators in a gallery. The content was theatrical, dramatic, and hugely symbolic, provoking either encouragement or dismay in its deadly conclusion.

By the end of the American Civil War the term 'sharpshooter' was in common currency thanks to Berdan's regiment. Not only had they altered the tactics of warfare, they had added to its lexicon. The word 'sharpshooter' has nothing to do with the Sharps rifle favoured by Berdan. In common with many things connected with long-distance shooting, it comes from Continental Europe. 'Sharpshooter' derives from 'Scharfschütze', which is German for 'marksman'.

The word 'sniper' was also slowly entering the vocabulary. It is thought to have originated with soldiers in the British Army in India around 1770, coming from one of their favourite pastimes, shooting at snipe – a small bird that flies swiftly and erratically. Shooting snipe on the wing would test even a very skilled marksman. The word gradually became better known as soldiers wrote home or returned to England, and by the end of the 1800s any long-range shooter operating within the British Empire would be referred to as a sniper. The first use of the word to mean someone who shoots from cover is dated to 1824 in the *Oxford English Dictionary*.

9

GOD AND THE MAUSER

The combination of the technology of the breech-loading rifle and the Minié ball, taken with the organisation of Hiram Berdan, had radically altered the combat capability of the infantryman. The rifle had become the deadliest and most versatile weapon on the battlefield and the rifleman need fear no one. When wars end, however, snipers become their forgotten combatants, and so it was after the American Civil War. A whole manual of tactics and field craft could have been written to accommodate this new branch of warfare; instead the snipers were allowed to fade into the background, with their expertise and wisdom left largely ungathered.

In the United States the sharpshooting regiments on both sides were disbanded and their members drifted back to their farms and their factories. Those who were particularly good shots often took the advice of Horace Greeley and went West. There, on the rapidly expanding frontier, they could earn a good living as buffalo hunters. One of the best documented examples of their skills came in 1874 at the battle of Adobe Walls in the Texas Panhandle, where a group of about thirty such hunters and local merchants was surrounded at a trading post by a band of several hundred Comanche. Calmly and methodically the buffalo men used their large-calibre Sharps rifles to put their attackers to flight. One hunter, Billy Dixon, took aim at a group of Comanche on a ridge near the trading post. He fired: one of them fell dead from his horse, the remainder fled. Later the shot was estimated at between 1,200 and 1,600 yards; even at the most conservative estimate it is a prodigious feat of marksmanship. Hardly surprising that the Comanche called the Sharps buffalo rifle 'the gun that shoots today and kills tomorrow'.

Although the specialist regiments were broken up, work continued to refine the technology of the rifle. With the development of a smokeless propellant in 1886 the way was almost clear for the arrival of the modern sniper. After nearly 100 years of technological innovation, by the beginning of the twentieth century the rifle had reached a point in its development where it was essentially the weapon that we use today. All of the problems – breech loading, distance and accuracy, smokeless propellant – had been overcome.

Improvements in ballistics gave a kick-start to new developments in optical science, specifically the telescopic sight. Anyone taking aim with an open-sighted rifle, thanks to the principle of parallax, can have a sharp image of the target or the weapon they are firing but not both. The solution to the problem lay at sea. The telescope had been around since the early seventeenth century, and sailors, who referred to it as a 'glass', were quick to seize on its possibilities. A lookout in the crow's nest with a glass enabled the fleet to see further; enemy ships could be spotted before they broached the horizon visible from the deck. In addition the telescope – the name came into use in the Royal Navy in 1774 – made it easier to read signals and generally improved communications between the ships of the line.

While he was fighting in the Crimea, Lieutenant Colonel Davidson of the 1st City of Edinburgh Rifle Volunteers had watched riflemen in the trenches during the siege of Sevastopol. The snipers were there to target Russian artillery crews, and Davidson watched as they worked in pairs, forming a highly effective partnership with one using a telescope to spot for the other. Whenever he saw a possible target the spotter told his partner to fire. Davidson realised that there would be potential if the man firing the gun could see over those distances on his own, so he came up with the Davidson sight, which was a modified telescope that could be mounted on any rifle.

By the end of the American Civil War what we might recognise now as the telescopic sight was becoming common. The early versions, fitted on target rifles, were little more than telescopes attached to the barrel of the gun. They gave a magnification of about one and a half times. Focusing was by means of an adjustable eyepiece; not terribly sophisticated but effective. By around 1880, the first proper telescopic sight had been developed in Austria and by the turn of the twentieth century, although very expensive, the telescopic sight was a common addition to big-game or target rifles. Equipped with such a gun, a hunter at the beginning of the twentieth century could see and fire accurately up to a range of a mile and a quarter.

By 1900 most of the world's military had rearmed their troops with this new generation of rifles and the sights that went with them. And at almost the same time one of the world's great armies was given a tragically costly demonstration of just how effective the modern rifle could be.

The Boers of South Africa were farmers. Descendants of Dutch settlers, they saw no reason why they should be ruled by the British and had gone to war with Britain in 1880 over their desire for independence. Britain had already recognised the autonomy of two Boer republics, the Orange Free State in 1852 and then the Transvaal Republic in 1854. But when the Transvaal became bankrupt in 1877 it was annexed by Britain and recolonised. By 1880 the Transvaal Boers were ready again to demand their independence. When London refused they made a unilateral declaration on 20 December 1880. A brief but bloody conflict ensued, which lasted until the following August. This was the First Boer War. In the course of its eight months, a well-organised, well-disciplined regular Army was soundly embarrassed by an irregular force of farmers.

The Boer farmer was born to the rifle in much the same way that the earlier American colonists or the Confederate sharpshooters had been. Carving an existence out of a harsh environment meant that they had to be able to pick up a rifle almost as soon as they could walk, and they had to learn to use it and use it well. If you were hunting springbok or more dangerous game with a muzzle-loading rifle you seldom had an opportunity for a second shot. If you were unable to conceal yourself properly then the game would not approach within range. Ammunition was scarce and very expensive: wasted shots simply could not be afforded.

The ivory hunter Charles Baldwin paid a generous tribute to the marksmanship of the Boer farmers from his own experience on the hunt.

> ... the day invariably wound up by target-shooting, at which the Dutch are great adepts. A yokeskey [wooden peg, part of an ox-yoke] at 100 yards, or a bottle, was frequently the mark, and sometimes the crack shots called for Eau de Cologne flasks, short, squab little things, no higher than a wine glass, and looking uncommonly small at 100 yards, which were, notwithstanding, frequently smashed.

One of the Boer commanders, General Wynand Malan, claimed that one of the best shots under his command, a man called Judge Hugo, could shoot the head off a korhaan – a small pheasant-like bird – at eighty paces.

The Boers, crack shots and expert horsemen, knew the terrain intimately. They were formidable opponents. The British troops stationed in South Africa, on the other hand, although professional, were not born to the horse, had never handled a weapon till they joined up, and the landscape was completely foreign to them. In addition they were led by men who were tactically naïve.

A great many men died needlessly in the American

Civil War because their generals did not appreciate the changes that the rifle and the Minié bullet presaged. Commanders on both sides were slow to realise how deadly the frontal attack had become. The British troops in the Transvaal were under the command of Major General Sir George Colley, who was governor of both Natal and the Transvaal. He too was unreceptive to the lessons of history and his men were condemned to pay the price.

Armed with Browning rifles, the Boers inflicted heavy damage on Colley's forces. The British were beaten in several significant encounters, almost always when they were attempting a frontal assault on Boers who were well armed, well motivated, and mobile. On 26 February 1881 Colley led 350 men on a night march to capture Majuba Hill. The Boer commander General Piet Joubert sent small squads of sharpshooters up the hill. Under heavy covering fire these Boer marksmen wiped out almost the entire British force. Colley himself was killed by a sniper's bullet. Less than a month later an armistice was signed and the First Boer War formally ended. Independence was restored to the Transvaal.

The First Boer War was a prelude to the much greater confrontation that erupted eighteen years later. The peace negotiated in 1881 was an uneasy one and the Boers still felt their autonomy was under threat, especially after the creation of Rhodesia, which was British-owned, to the north and the burgeoning prosperity and ambitions of the British colonies of Natal and Cape Colony. More and more British workers were flooding towards the gold mines of the Transvaal to seek their fortune. The empire-builder

Cecil Rhodes was behind an attempt in 1895 to stir up these foreign workers against the Boers. The insurrection was suppressed, but it alerted the Boers, who began to prepare for another war.

During the First Boer War most farms possessed only one gun. In 1896 the Boers started importing large numbers of German-made Mauser rifles so that they could mobilise larger numbers of men. The Mausers were state-of-the-art, part of a new generation of weapons being turned out by companies in Europe and America such as Mannlicher, Enfield and Remington. The Boer leader Jan Smuts claimed that he put his faith in two things: 'God and the Mauser'.

The 8mm Mauser was accurate at a range of 1,200 yards for aimed shots and almost twice that distance for volley fire. Its bolt-action allowed for a great rate of fire. The clip-fed magazine held smokeless cartridges. The ammunition was new too. The bullets had hard metal jackets and were larger than the Minié ball – about three-tenths of an inch in diameter, or .30 calibre. Smokeless powder burned quicker and more efficiently than the old black powder, so these new, hard, bullets travelled further and faster than their predecessors. They hit their target with much greater force and did much more damage to whatever they encountered.

With their new rifles the Boers were ready for combat again when the Second Boer War began in October 1899. In the process they would also add a new word to the combat vocabulary.

Commando is a Portuguese word that has now become synonymous with highly effective guerrilla troops. In both the Orange Free State and the Transvaal there was a legal requirement for men to 'go on commando'. This meant that when they were called for military service they were required to provide their own horse, rifle and ammunition, and their own food for a week. The commandos operated on a democratic structure, with officers elected from within the men rather than appointed by a central command. They were fast and clever, they were crack shots, and the British Army had no effective means of dealing with them.

In October and November of 1899 the Boers launched several lightning manoeuvres against major towns. They attacked Mafeking and Kimberley as well as laying siege to Ladysmith. Vital to the attack on Ladysmith was Spion Kop, a flat-topped mountain that overlooks the town. The Boers had taken control and dug themselves in. It was imperative that the British retake the hill if the Boers' hold on Ladysmith was to be broken. In January 1900 the British attempted a surprise night attack. If they could clear the Boer guns from the summit then the way would be clear for the British troops to advance to relieve Ladysmith, some twenty miles away.

Hundreds of British soldiers were killed in the day-long Battle of Spion Kop. The mountain top was referred to as 'an acre of murder'. The minute they got to the top of the hill the British troops were exposed to the withering fire of Boer marksmen. The British claim that 300 men died; other observers suggest the figure was closer to 1,000. Among the victims of the Boer snipers was the British commander, General E.R.P. Woodgate, shot through the head at Aloe Knoll. One grim testament to the skill of the Boer marksmen is that after the battle seventy-five bodies were found in a single trench. They had all died from single bullet wounds to the head, victims of Boer snipers with rapid-fire Mausers. The British had thought they were secure, but their trenches were too shallow to provide cover from the Boers' deadly fusillade.

The British also had new weapons. By the time of the Second Boer War they had upgraded their rifles for Lee-Enfields, good guns that could hold a ten-round magazine. They did however use black powder where the Boer Mausers were using smokeless ammunition, which meant that the Boers could continue to fire from cover without giving themselves away. The Boer commando Deneys Reitz claims the British were also tactically naïve: 'Time after time I saw soldiers looking over their defences to fire and time after time I heard the thud of a bullet finding its mark and could see the unfortunate man fall back out of sight, killed or wounded.'

The British were at a loss as to how to deal with this new menace. Where the British still fought in twenty-five-man units, the Boer commandos were flexible and ranged around the battlefield where they pleased. Unlike the British, they were not prepared to stand in the open and exchange fire; they would go to ground and conceal themselves, or fire from trenches. British officers also frequently complained that even when they suspected the Boers were present they could not be seen: the commandos were invariably bearded and wore large floppy hats that put the rest of their face in shadow. With their faces completely obscured and no hint of skin showing, this created an accidental camouflage that served them well, especially when firing from shade or at night.

The Boers' biggest successes were in the early days of the war. The British slowly began to gain the upper hand through sheer weight of numbers, but even when the tide was turning against them the Boer commandos continued to be effective in sniping and harassing British columns and convoys. Eventually the British started sniping back. It was an ad hoc response, not a formally instituted tactic. However it was not uncommon for the British to rely on the local knowledge of the South African Police or the Rhodesian Regiment as counter-snipers. These men would creep out from their own lines before dawn and, waiting until the sun was behind them, they would then snipe at Boer artillery crews before making their way back to their own lines under cover of darkness.

The Second Boer War ended with a peace agreement at Vereeniging in May 1902. It created a new South Africa in which the minority British and Boers shared power over the majority non-white population. Britain lost almost six thousand troops in combat during the war, while the Boers lost 4,000. A further 30,000 men, women and children from both sides died of malnutrition or disease.

The losses suffered against the Boer riflemen caused public embarrassment for the British Army. There was a flood of charitable donations to set up programmes to allow soldiers to be trained at shooting moving targets. After the war Lord Roberts, the British commander, gave his support to a campaign to get working men to take up shooting. The aim was to improve marksmanship and turn the country into a nation where there was a 'rifle in every cottage'.

Britain had learned a costly military lesson in South Africa. It knew that its battlefield tactics were no longer relevant to the modern age. It knew also that its standards of shooting were equally unacceptable. Both issues would be urgently addressed, but not in time to prevent another, more tragic, education.

10

OFFICERS AND GENTLEMEN

There was no more dangerous place in the world to be in the spring of 1915 than in an Allied trench on the front line of the Western Front. Life expectancy could be measured in minutes and frequently was. The unwary newcomer, sense and reason lost to an adrenaline rush, was inclined to stick his head over the lip of the trench to glance across towards the German lines. The same naïveté evident in South Africa brought an invariably similar outlook. That reckless glance was often all it took for the newcomer's first look at no man's land to be his last. Even hardened veterans were not spared, and more than one seasoned campaigner paid the price for showing off to raw recruits.

The German sniper reigned supreme on the Western Front in the first stage of the First World War: any careless movement by a British soldier was likely to be followed by a swift and fatal resolution. A shot would ring out, followed almost simultaneously by the telltale ping of a bullet hitting and penetrating a steel helmet, leaving the unwary soldier dead at the feet of his shocked and demoralised comrades in arms.

One of the pivotal areas was Neuve-Chapelle, on the road between Béthune and Armentières. In February and March of 1915 the area became what was later grimly described as a sniper's paradise. Most of the town's houses were still standing and provided perfect cover for German marksmen. Armed with hunting rifles and using telescopic sights, they picked off the men of the British 8th Division almost at will. Whatever they did seemed to draw the sniper's fire. A Lieutenant Shingleton of the British 7th Divisional Artillery noted that even the most innocent of acts could invite catastrophic consequences:

> . . . to be seen standing in the open meant instant death from a sniper's bullet. It was even dangerous to wear a luminous wrist watch, for the light from same could be seen a long way off in the pitch darkness; a few snipers' bullets had whistled past us and the officer in charge suggested that my luminous watch was the cause whereupon I promptly put it in my pocket.

It's hard to imagine the terror that must have been induced by an enemy who could zero in on a wristwatch at a distance of 400 yards on a moonless night. It was estimated that it took barely three seconds for a German sniper to spot, sight, and kill his target. If luminous watches were a risk then smoking in the trenches posed all sorts of health hazards other than those that would be discovered later. Experienced soldiers never shared a match. They knew that a sniper would be attracted by the flare of the match, set his sights on the second person getting a light, and shoot on the third. Three lights off the same match was considered unlucky, a superstition that persisted into civilian life for those who had seen service on the Western Front.

The war poet Siegfried Sassoon almost met his end at the hands of a German sniper. Sassoon was lucky enough to survive and sufficiently literate to give a vivid account of his experience:

I thought what a queer business it all was, and then decided to take a peep at the surrounding country. No sooner had I popped my head out of the sap than I received what seemed like a tremendous blow in the back – between my shoulders. My first notion was that I had been hit by a bomb from behind. What had really happened was that I had been sniped from in front. Anyhow my attitude towards life and the war had been instantly and completely altered for the worse.

Sassoon was a very fortunate man. The bullet passed clean through his chest at shoulder height without hitting any internal organs. Since the German snipers favoured a clean head shot we can perhaps surmise that Sassoon was so far away from the shooter that the bullet had arced downwards in flight and hit below its intended target.

Many men did not share Sassoon's good fortune and met tragic, pointless and untimely deaths on the Western Front. Private Nobby Clarke, a veteran of the Western Front, summed up the experience commendably succinctly. 'Anyone who had a month in the trenches in March 1915 should have a medal as big as a frying pan,' he said.

The First World War was not intended as a war of trenches. They arose from the failure of the Schlieffen Plan, which had proposed a swift conquest of France before Russia had time to mobilise its huge but under-equipped Army against Germany. The plan called for France to be overrun by German troops pouring into the country through neutral Belgium. The attack was launched on 18 August 1914, and historians argue still about why it failed. France simply refused to be overrun. Together with the British Expeditionary Force the French Army made a stand on the Marne, north of Paris. With the Germans within sight of the French capital, they were turned back by a superbly disciplined display of volley fire. Tactics that had served the British poorly in South Africa were used to great effect here. An enemy advancing en masse rather than the Boers' mobile horseback units was much easier to deal with. British troops firing at a rate of fifteen rounds a minute – the 'mad minute' – winnowed the ranks of the advancing German forces. A controlled display of volley fire and movement saw the British eventually halt the German advance in what became known as the Miracle of the Marne.

Very quickly the conflict settled down into a grinding war of attrition, as each side decided that what they had they would hold. The most technologically advanced conflict the world had seen became mired in the static patterns of the pre-industrial age. The territorial lines that had been established in the early part of the war remained fairly constant for lengthy periods as the world prepared itself for its first whole-sale experience of trench warfare.

Trench warfare is in many ways an extension of siege warfare. It was warfare by relentless battery, charge and counter-charge, across ever more impassable fields. It was not tactically sophisticated. The cost in human life was appalling.

It was by no means new. It had been seen in the American Civil War and before that at the siege of Sevastopol in the Crimea. Britain had learned from bitter experience in South Africa just how difficult it could be to storm a trench. The Boers had proved that by making a line long enough it would be next to impossible to turn the flanks, especially if the trench could be filled with enough men to cover the front. Developments in weapons technology meant that in 1915 a line could be held with only two men per yard, as opposed to the five men who had been required fifty years previously. Neither side in the First World War lacked manpower, and eventually the fortifications stretched over miles of terrain from the Swiss border to the North Sea.

There is no doubt that the war in the trenches was fought on terms dictated by the Germans. They dug theirs on forward-facing slopes in well-drained ground. They also had the ordnance in the form of howitzers and heavy guns to fight a trench war. The Boer War had taught the British the efficacy of the 'creeping barrage' in dealing with an entrenched enemy. Artillery could provide devastating cover for an infantry advance that would allow them to close on a hopefully demoralised enemy. The commander in chief of the British Expeditionary Force, Sir John

French, requested heavy siege guns to respond but they were slow in coming. The French had no howitzers at all and had to take heavy guns from existing fortifications and ship them to the front line.

Private E.F. Clarke, known to his mates as 'Nobby', enlisted in the 1st and 2nd Northants Regiment on 13 January 1914. Clarke came from Scotland Road in Market Harborough in Leicester and was barely eighteen years old when he signed up. He became a runner, an incredibly dangerous operation that involved ferrying messages back and forth between the lines, frequently under heavy fire. His view of life on the Western Front was entirely typical of the hundreds of thousands of Allied soldiers who fought there. It was like nothing they had ever seen or been prepared for.

> We had never trained for trench warfare. Moreover they had a 10–1 advantage in machine guns; and snipers with telescopic sights – we had none ... their trenches were mostly on high ground with about six yards of barbed wire in front. Our trenches were a foot deep in mud and water.

In 1914 the trench was a relatively simple affair, nothing more elaborate than a hole in the ground deep enough and long enough to shelter a few men. As the war progressed and the combatants swelled in numbers, so they became more complex. The trench had become more than simply a hole in the ground: it had become a way of life for those occupying it.

Traditionally both sides had a front line in which they would station troops required to go over the top and do the fighting. This could in practice be a series of two or three trenches with others in between to aid communications. Behind this there would be a support trench, ideally with some kind of hill or ridge between the two allowing your forces to regroup for a counter-attack in the event of the main trench being overrun. The Germans frequently had a third line behind this again. As well as the communications trenches, another ran all the way from the front line to the rear to allow casualties to be evacuated to a field station.

The trench was extended so that it was deep enough to allow a man in full combat gear to walk upright without being spotted and shot at by the enemy. The British, unable often to dig on well-drained higher ground, had serious problems with water seepage through the clay soil. Duckboards had to be laid along the bottom of the trench to prevent soldiers sinking up to their ankles or beyond in the cloying mud. About eighteen inches above the duckboards on the side facing the enemy there was a fire-step on which the soldier could stand to allow him to fire out into no man's land. A stout earthern berm on either side of the trench offered some protection. The berm on the front was the parapet, that at the back was known as the parados. British trenches were also dug in a Greek key design – a flattened sawtooth – which limited the damage if a shell landed in the trench. Typically both sides strung coils of brutal barbed wire to slow down advancing enemies and turn them into easier targets.

Troops were rotated in and out of the front line every eight hours or so. The stress must have been almost unbearable as they sat helplessly waiting to be either shelled or go over the top and face the enemy machine-gun fire. Discipline was strict. No one was allowed to put their head over the parapet and only snipers were allowed to fire in the daytime. It was, as Nobby Clarke recalls, 'more or less standing in a coffin with the lid open'.

Nowhere was the lid of that coffin more likely to be slammed shut than in the Allied lines in the early months of 1915. The trench war had become a snipers' war and the Germans were pre-eminent. The Jäger tradition meant the German sniper reigned supreme: no man's land was his killing field.

The accuracy of the German snipers had the Allied forces thoroughly demoralised. On any day in the early months of 1915 a battalion – 1,007 men at full strength – in even a quiet sector of the line could be certain of losing up to eighteen men to sniper fire. The effect on the morale of the men standing-to in the trenches was devastating.

The Germans had the run of the battlefield. So confident were they of remaining undetected that they would even lie out in the middle of no man's land using the bodies of their dead as cover. They shot from distances of between 200 and 400 yards, using rifles with muzzle velocities of around 3,000 feet per second – more than twice the speed of the deadly Mauser of twenty years earlier. Their intended target was dead before he had a chance to hear the shot that killed him. The tiny entry wound in front became an exit wound as big as a fist, spilling brains, bones and brawn everywhere. Intensely demoralising, it was a sight to sicken even the hardest of soldiers. There is some evidence that the Germans used soft-nosed 'dumdum' bullets which spread on entry, creating an even more devastating exit wound.

There was an obvious and possibly criminal tactical failure on the part of the British Expeditionary Force to address the problem of German snipers. It was perhaps, as General Lord Horne admitted later, more than anything a failure in imagination: 'Possibly Germany was more quick to initiate new methods to meet prevailing conditions. Certainly we were slow to adapt, indeed our souls abhorred anything un-sportsmanlike'. This legendary British sense of fair play and decency certainly inhibited their tactics in the early part of the war. Horne pointed out that, for example, had it been up to the British we would never have seen the introduction of gas to the battlefield. Gas was definitely unsporting, and so too was sniping. But the losses were too great to ignore or sustain. Something had to be done and, as Horne says, 'once we set our minds to something we quickly caught on'.

Horne was not strictly correct. The British would eventually win the sniper war, but only after a labori-ously slow struggle to adapt.

Major H. Hesketh-Pritchard – known to his friends as 'Hex' – was an infantry officer who had been a big-game hunter in Africa. He was aghast at how easily the British were being picked off by German snipers. Hesketh-Pritchard had brought his own hunting rifles to France, and would frequently go out into no man's land to stalk the enemy. But his were solitary, occasional successes only.

In 1915 the German Army had 20,000 high-quality

telescopic sights fashioned by leading manufac-
turers such as Zeiss. In addition they had some of
the best weapons in the world on which to place
them. Additionally a German nobleman, the Duke of
Ratibor, had done the Kaiser an incalculable service
by starting an appeal for all the sporting rifles in
Germany to be collected up and sent to the Front.
Britain by comparison had next to none. Individual
officers like Hesketh-Pritchard who had been big-
game hunters, or those who had been ghillies or deer-
stalkers, brought their own Mausers or Mannlichers.
Other than that the British soldier had to rely on the
standard-issue Lee-Enfield rifle.

Germany used its snipers zonally. Each shooter
would patrol a half-mile stretch of the line selecting
targets of opportunity pointed out by sentries who
acted as ad hoc spotters. A single German sniper
could be responsible for thirty to forty British casualties before he was either shot or, more likely, moved
on to another part of the line. They also managed to
shoot a disproportionately high ratio of officers, which
had its own effect on the command structure. The
Germans later revealed they could tell the difference
between officers and other ranks because 'the legs of
officers are thinner than the legs of enlisted men'. It
is pointless but tragic to speculate how many men
died because of the vanity of their immaculately cut
riding breeches.

Britain had no adequate response. Those who
coped best were units such as the Canadians, who
had hunting experience, or those commanded by
officers who had been big-game hunters or deer-
stalkers. They realised that even the most basic of

counter-measures such as rags hung on barbed wire to distract the sniper's peripheral vision could save lives. The British troops generally made no effort at concealment. Where the parapet of the German trenches was broken and irregular to afford cover and protection, the British prided themselves on a parapet that was almost maniacally beaten flat to provide neither concealment nor security.

Early attempts at counter-sniping were crude but occasionally effective. A concerted attack by rifle grenade was often enough to destroy large numbers of German observation telescopes, and in the process purchase some respite from the sniping. Other counter-measures were a little more fanciful. Nobby Clarke had graduated from being a runner to being a bomber, an occupation every bit as dangerous. Bombers would crawl out into no man's land with several rudimentary grenades. Armed only with a knife they lay, bomb in hand, until another team member lit the fuse, allowing them to throw it.

On his way back from one of the nocturnal bombing missions in March 1915, Clarke became the Northants Regiment's first 'sniper' by accident. He met someone he knew, a Boer War veteran considered to be the father of the regiment, and was shown how to use the sort of contraption Heath Robinson would have delighted in.

'The gadget was a kind of tripod and you lay the rifle on top of it,' Clarke recalled. 'There was a looking glass in the front and another at the back. There was a piece of wire, you put the loop round the trigger, and then the other end on a lever in the front. To use it you had to lay [it] flat on the parados . . .'

Cumbersome and complicated, this gun could not be moved to follow an enemy target, but would instead have to wait until a target came into its field of fire. It was still the best the British were able to come up with: small comfort for the men in the trenches. 'If this is the best sniping weapon our men in England can turn out then we've had it,' Clarke noted bitterly.

Other soldiers shared his opinion, especially since once the weapon was fired its position was transparent to the Germans, who would retaliate with vehemence. This sort of action was known to the British troops as a 'hate'. It provoked a heavy concentration of artillery, rifle, pistol and grenade fire on a single point on the line. A hate was something to be avoided whenever possible. The sniper became a curse to his own troops, a pariah. But as Clarke discovered, there were compensations: being a sniper improved his rations quite considerably. Wherever he took his gun on the line, soldiers would attempt to bribe him with tea and cigarettes to take it elsewhere for fear of drawing German fire.

In contrast to the massive scale of the slaughter caused by bombardment and massed charges, the sniper duel was personalised. It transformed the conflict into a war of individuals.

Britain needed a more effective counter-measure against German snipers. What few resources they had were being squandered. Telescopic rifles were simply handed out as part of trench stores, with no attempt to give them to specialists. Hesketh-Pritchard reveals how, on one occasion, he saw one soldier patently unskilled in the use of the telescopic rifle simply sticking his head above the parapet from time to time to take pot-shots at the enemy.

The serious business of matching the German snipers was taken up in earnest by Hesketh-Pritchard. He believed telescopic rifles were the key: he felt they were unsporting but doing without would maintain the German advantage. The British telescopic sights were very poor. Around 80 per cent were out of alignment, and there was no instruction on how to correct them. If a sight is out by as little as one-hundredth of an inch, then that can mean a deviation of 9 inches at 100 yards, 18 inches at 200 yards, and 54 inches at 600 yards.

Hesketh-Pritchard told his commanding officer, Lieutenant-Colonel A.G. Stuart of the 40th Pathans, that he wanted to train British troops in the use of the telescopic sight, basic camouflage, and shooting skills. He wanted to set up a sniper school to create a unit of specialists to counteract the German threat.

Hesketh-Pritchard found two important allies to his cause. Sir Charles Monro, Commander of the British 3rd Army, and the writer John Buchan, then a war correspondent for *The Times*, believed passionately in what Hesketh-Pritchard was trying to achieve. Buchan returned to London and used his society connections to set up a fund to buy rifles. The fund attracted several prominent supporters and the Army was ultimately able to acquire fifty-two rifles. Buchan was also able to provide Ross telescopes for the spotters, which offered twenty times magnification and were far superior to the standard-issue military field glasses.

Even with this support, Hesketh-Pritchard's scheme almost foundered because of bureaucratic intransigence. He and a colleague, Colonel Langford Lloyd, who had established a telescopic sight school, jointly wrote a training pamphlet, but it was never published because there was a change of personnel and prevailing opinion at High Command, typical of the Army's attitude at the time. He was told on several occasions that a sniping school could not be established because there was nothing in the Army establishment to provide for it. Finally Hesketh-Pritchard had to accept a demotion in rank to infantry captain, with an attendant cut in pay, to get his idea off the ground. He then went unpaid for eight months, but didn't mention it for fear of causing a fuss at GHQ that might have led to his operation being discovered and then cancelled.

The first task that Hesketh-Pritchard and his loyal cadre of like-minded officers set themselves was to make sure that the telescopic rifles already in France were up to the job. They visited brigades up and down the line to show what sort of improvement could be achieved with only the slightest modification and training. They would turn up at a brigade and fire the rifle as it was, invariably missing the target by some distance. They would then zero the sight and fire again, dead on target. Hesketh-Pritchard himself says that he and his colleagues were seen as 'performing animals', and indeed it must have seemed like Buffalo Bill's Wild West show come to town when they pitched up to provide much-needed diversion. Resistance from the brigade commanders was considerable. On one occasion Hesketh-Pritchard and his men were told: 'You are not here officially and any Germans you may kill or cause to be killed are of course only unofficially dead.'

Unofficial or not, German snipers could finally be targeted. For Hesketh-Pritchard sniping was the purest form of shooting. Unlike with big-game hunting there was no chance of a sighting shot, because the target would go to ground. In any event the target was seldom willing to show more of itself than a cap badge or, at best, half a head. It was, according to Hesketh-Pritchard, 'the highest and most difficult of all forms of rifle shooting'.

After the conflict, when he came to write his memoirs, Hesketh-Pritchard divided his snipers' war into four distinct phases, the first phase lasting until the spring of 1915, during which the Germans had almost complete mastery of the field.

Hesketh-Pritchard finally established the first sniping school at Béthune near Armentières in 1915. This was known as the First Army SOS School – 'SOS' standing for Sniping, Observation and Scouting – and this was quickly followed by the establishment of the Second Army School of Sniping.

Candidates at the school took a seventeen-day course covering the basics of rifle and sight maintenance, shooting, observation, camouflage and scouting. The weaponry may have changed, but the course as it is taught now does not differ too much in principle from the one laid down at Béthune. There were, according to Hesketh-Pritchard, three components to the art of sniping: finding your mark, defining your mark, hitting your mark. He believed that anyone could be trained to be a sniper. He also believed that unlike the Germans, British snipers should work in two-man teams of a sniper and a spotter. More importantly, they should switch roles at regular intervals, as often as every twenty minutes. Although a keen shot himself, Hesketh-Pritchard actually believed the spotter had the place of honour in the partnership, since his was the more stressful occupation.

The school at Béthune paid almost instant dividends, and the training began to filter down to men like Nobby Clarke. He received one of four telescopic rifles allocated to his unit, and the difference was immediately obvious. Clarke believed it was now almost impossible to miss. 'You could see if a soldier wanted a shave at 400 yards,' he said, which, as he reasoned, would be why the Germans had been so successful in picking off British officers, since they were allowed facial hair.

It would be rare for a British sniper to operate at a range any greater than 400 yards. 'The chances of hitting a German head at 600 yards . . . if there is any wind blowing at all, are not great,' said Hesketh-Pritchard. 'We therefore . . . never went back further than 400 yards, and our greatest difficulty was to teach the snipers to appreciate the strength of the wind.'

Wind was one of several factors that Hesketh-Pritchard had to teach his snipers to contend with. Heat haze could also be a problem. There was also the question of drift, the tendency of a bullet to veer either left or right depending on the direction of the rifling of the barrel, about an inch for every hundred yards. At a distance of 400 yards the sniper would have to aim his shot four inches to the right or left of his target.

Even so, British soldiers proved to be generally quick learners. According to Nobby Clarke hundreds of shots were fired, most with deadly accuracy, but the snipers only ever reported the number of shots they used, not the number of kills. 'Never be bluffed if a sniper swanks how many he has killed,' cautioned Clarke. 'None of our lads did. We knew it was almost impossible to miss but unless the enemy is shot in the open and lies for a while you could not be certain.' Clarke even made a morbid joke of it when a particularly overbearing Sergeant Major demanded figures on the German dead. 'Sir,' Clarke told him, 'we are waiting for the Germans to send their report in.'

So followed the second phase of Hesketh-Pritchard's sniper war – to isolate and pick off German snipers. As German casualties mounted British morale improved. In addition British snipers performed another vital function: intelligence gathering. The SOS School at Béthune had placed as much emphasis on scouting as it had on shooting, and the snipers it turned out were an invaluable source of information on German activity. In the summer of 1915, for example, just before the Battle of Loos, Nobby Clarke recalls that he and his fellow snipers were employed almost full-time on intelligence gathering. He remembers they had to 'look at every nook and cranny behind Gerry's line'.

Hesketh-Pritchard's 'secret weapon' in the campaign against the German snipers was the Lovat Scouts. This was a crack unit of around 200 men who had all been Highland gamekeepers – ghillies as they were known in Scotland. Their powers of observation were legendary. Years of work stalking deer and other game in the remote hills and glens had given them an unrivalled visual acuity. The Lovat Scouts could see things with the naked eye that trained snipers could not spot even with a telescopic sight. Hesketh-Pritchard loved his Scouts and his trust in them was implicit. 'If they reported a thing, then the thing was as they reported it,' he said simply.

The Lovat Scouts also taught British snipers the black arts of concealment and camouflage, without which they could not have plied their trade with anything like the effectiveness they did. 'Snipers should be artists in camouflage . . . if they have any desire to continue their existence . . .' recalled W. Carson Catron, himself a former sniper. 'Hands, face and neck were tinted with a mixture of water and mud and allowed to dry on. A grass sod with soil removed is placed over the sniper's head interlaced with longer grasses.'

In addition to this the Lovat Scouts also introduced the ghillie suit that is now a staple of the sniper's uniform. The ghillie suit was a loose-fitting robe, usually hessian, into which could be stuck leaves, branches, soil, material and anything else that was required to break up the shape of the sniper. Applied properly, a ghillie suit should be impossible to spot until you are standing almost on top of the sniper. Observation could then be carried out without risk, sometimes with spectacular results.

Hesketh-Pritchard himself reported on one remarkable incident involving snipers from the Royal

Warwickshire Regiment who had observed a tortoise-shell cat sunning itself on the same part of a trench for several days in a row. Their suspicions were aroused because the particular section of the German line was supposed to be unoccupied. Their observations were reported to headquarters, where a sharp intelligence officer reasoned that the cat was there because it was being fed. He believed the cat was being encouraged by German officers keen to keep it around to deter the rats swarming throughout the trenches on both sides. Aerial pictures were ordered, a German head-quarters was discovered, artillery was called in, and the command post was destroyed. All because a cat was spotted stretching in the sun.

Hesketh-Pritchard's war entered its third phase once the successful efforts of the British snipers had driven the Germans back to ground. No longer bold enough to lie in no man's land to take their shots, they now had to be flushed out. The campaign of sniping and counter-sniping became as much an intellectual exercise as a military one. It was a deadly battle of wits.

British soldiers had learned some of the brutal lessons of the Boer War. The watchword in that campaign had been 'keep well clear of officers and white rocks'. The Boers, like the Germans on the Western Front, seemed to have an infallible instinct for spotting British officers. The white rocks had been deliberately placed to act as range markers. The men trained by Hesketh-Pritchard adopted similar tactics. Laziness and predictability could be the death of many a soldier. Their routine and their habits gave them away.

'Without thinking they make a nice little niche in between two sandbags and they settle their rifle in that niche,' recalled Carson Catron of the sort of behaviour that got sentries killed. 'He stands down, another sentry takes his place and puts his rifle into the same niche and stands behind it. He looks for the niche since it saves him the bother of making one for himself. This is one reason why snipers study tops of trenches.'

Snipers on both sides protected themselves by inserting steel plate between the layers of sandbags of their sniper posts. The plate had a covered hole cut into it to allow the sniper to pull back the cover of the hole, slide his gun through, and take his shot while still protected by armour. The loophole, as it became known, saved a lot of lives, including that of Frederick Sleath of the 51st Highland Division in a sniping duel at Ypres. He was taking aim at a German sniper when the German fired first and the bullet thudded into the steel plate barely two inches from where Sleath was shooting. He calmly resighted, took aim, and killed his quarry.

The use of loopholes initially exposed a design flaw in the British sniper rifle. The original telescopic sights were mounted off-set on the side of the rifle. Hesketh-Pritchard himself had first-hand experience of this:

There was a sniper beside me who had one of my rifles, a Mauser which had a telescopic sight on top, and with which he was able to fire through the loophole . . . a working party of Germans appeared . . . they had but a few yards to go to regain their trench. The sniper who was next to me got off a shot, but two of the snipers armed with the government weapons . . . who were waiting at loopholes found that neither of them could bring their rifles to bear at the extreme angle at which the Germans were disappearing.

Once identified, this fault was quickly remedied. Concealment was vital, and every precaution was taken not to give away the position of the loophole. It was drummed into the troops for example that they should never open the loophole with the sun behind them. Additionally, loopholes were always opened from the side. Standard operating procedure was to open the loophole, expose a cap badge, and if there was no shot after 75 seconds then it was deemed safe enough for the sniper to take up his position. Once he was securely and secretly in place, the next task was to discover where the German snipers were.

Lieutenant Frank Glover joined the 1st Battalion of the East Surrey Regiment in September 1916, when he was just over eighteen. He was trained at the sniper school at Linghem, where he met Hesketh-Pritchard and seemed to be impressed with his fame as an author and sportsman. The hunter turned sniper had been an amateur cricketer of some note with Hampshire before the war.

One of Glover's jobs was to collate the sniper reports for his unit. The entries noted the usual sightings of Germans seen passing gaps in the trenches, with particular attention to cap types, colours of cap bands, and the appearance and number of German periscopes. These periscopes, along with observation mirrors, were a prime source of spotting what the other side was doing. Glover and his men engaged in a daily routine of spotting and smashing periscopes and mirrors. The rest of the entries paint a fascinating picture of a sniper's daily routine on the front line.

Jan 23, 1917
Heat shimmer was observed during the morning as from a smokeless fire in the German front line trench at . . . A loop hole plate is visible here and the parapet for an area of about 5 square yards was clear of snow as if it formed the roof to a shallow heated dugout.

Jan 24, 1917 1045am
The head of a German wearing a blue cap with a black peak was seen looking over front line trench at . . . Sniper fired and observer reported a hit.

Jan 25, 1917 1110am
A German, axe in hand, got on the parapet of their second line and, crouching down, proceeded to chop at something. Our sniper fired and the man dropped, rolling over into the trench.

Feb 23, 1917
A man in a peaked cap who fired a shot over the parapet of Canal Trench was fired at and hit by our sniper.

As the British became more skilled marksmen, the Germans were less and less willing to expose themselves to any risk. More sophisticated methods of bringing them into the open had to be tried.

Hesketh-Pritchard devised the ploy of constructing papier mâché heads to be stuck on poles and raised above the parapet. If the Germans shot at these dummy heads, a quick examination of the angle between the entry hole and the exit hole would reveal the shooter's position. When the Germans figured out this ruse, the British became more subtle still. A lit cigarette was placed in the dummy's mouth and a rubber tube run down into the trench so that the cigarette could be 'smoked' by a sniper in the trench. When the sniper's smoke was rudely interrupted the spotter could work out from where the German was shooting.

The modern sniper is used in many cases as a tactical shield, moving out in advance of his unit to provide protection and cover. This was a technique developed in what came to be the fourth and final phase of Hesketh-Pritchard's sniper campaign, in which the British snipers were used during the great battles. They would for example go out in front of the men when an enemy trench had been overrun to provide covering fire while the remainder of their troops poured into the trench. They had always been used to shoot out enemy periscopes to prevent observation. This practice was modified to take out enemy machine guns, either by shooting the crew or with a well-aimed shot to destroy the breech-block. The latter was the preferred option, since a machine gun was harder to replace than a gun crew.

The *London Gazette* of 16 September 1918 notes that Lieutenant Frank Glover of the East Surrey Regiment had been awarded the Military Cross for 'conspicuous gallantry and devotion to duty' in an engagement which had taken place in May of that year. The citation says that Glover 'worked day and night to secure information vital to its success. During the attack he advanced to the objective with the first wave, making a complete tour of the captured line and returned under heavy fire with a report to battalion headquarters.'

Only months later Glover won a Bar to add to his Military Cross in another action which took place west of Bapaume near the Somme in August 1918. This engagement perfectly illustrates the tactical importance of a sniper unit. The citation says that Glover 'went up with the leading wave – when it was held up he organised flanking parties which he moved at the same time as he delivered a frontal attack capturing 12 machine guns and 70 prisoners'.

This action does seem more intense than the first one. Glover's diary recalls how their 'every movement is sniped'. In a report written in the thick of the fighting Glover admits to having lost all trace of his men but takes comfort from the fact that he still has some snipers with him. In his own dispatches of the incident he gives credit to his Sergeant, a J. Lord, for rushing out under fire to pick off enemy machine-gun crews. Sergeant Lord then 'led the assault by his personal bearing and total disregard of danger'.

Also mentioned in dispatches of the incident is a Private Manning, for conspicuous gallantry in what reads like a textbook example of how a sniper should conduct himself. Glover reported:

> When the infantry were held up during their advance by heavy fire from a trench lined with enemy machine guns, this man crawled forward to a small mound where undisturbed by the heavy fire directed upon him he proceeded to pick off the enemy machine gunners thus largely assisting our men to gain fire superiority which eventually enabled the trench to be carried. When the final objective had been reached he again pushed forward to a vantage point from which he did great execution among the enemy with his rifle. As a sniper his coolness and devotion to duty at all times are beyond praise while he has personally accounted for large numbers of the enemy.

Three months later Armistice was declared and the First World War was over. There are no separate statistics that detail the ways in which soldiers died, no way of working out how many British troops were killed by German snipers. There is however no doubt that thanks to the efforts of Hesketh-Pritchard and a group of far-sighted senior officers the number was much smaller than it might have been. In contrast, in the first three months of 1918 no fewer than 387 Germans were confirmed killed by snipers of the British 38th Division alone.

By the end of the war the British Army had an elite group of soldiers who were trained marksmen, scouts and intelligence gatherers. Despite an obvious and incontrovertible display of their combat effectiveness there was still an unwillingness to commit to a permanent sniper force. The feeling still prevailed that this was not how wars should be fought or battles won. As General Lord Horne had somewhat archaically pointed out in 1915, there was still a notion that war should somehow be sporting, embodying the virtues of the playing fields of Eton. The prevailing view in the military establishment was that there was nothing sporting about sniping, it was simply unethical. The sniper was a soldier of expedience: coveted in wartime, spurned in peacetime.

Ultimately those who seized the moral high ground prevailed and the British sniper schools were disbanded. Hesketh-Pritchard was not there to lend his advocacy in any move to retain them. He died in 1923. The rejection of so much hard-won experience was a shortsighted and misguided decision. It would be barely twenty years before the British military establishment came to regret it.

11

A BREED APART

At the conclusion of the First World War snipers were specially trained marksmen who killed not in the heat of battle but with premeditation. They stalked their prey with the same skill and finesse with which many of them had hunted wild game before the war. As a weapon, the sniper was as much psychological as tactical. The presence of a sniper could induce panic or paralysis in a group of enemy soldiers.

It was not merely their unique expertise that set them apart. The sniper was excused normal duties, allowed to roam wherever he wanted in search of his quarry, answerable to no one but himself. Part of the attraction of being a sniper was the ability to control your own fate. Temperament was crucial. Hesketh-Pritchard found that there were some candidates who were marvellous marksmen but just could not come to terms with the qualities needed to stalk and kill another man. The snipers were the first specialised soldiers who had to possess a particular psychological aptitude. Mental strength was as important as physical competency. A hunting rifle and a telescopic sight meant a man could shoot almost anything he could see. Whether or not he would or should shoot was a different matter entirely.

Very early on, Hesketh-Pritchard realised he did not want hotheads, drinkers or smokers as snipers because these men might not be able to overcome their cravings so as to remain concealed for the duration until the kill shot. He sought out even-tempered, physically fit men who could do an unpleasant but vital job without any obvious qualms. These rudimentary standards introduced by Hesketh-Pritchard would be refined over the years: the modern sniper is subject to intense psychological scrutiny and profiling.

Attitudes to the sniper were to say the least ambivalent. W. Carson Catron recalls: 'They were hated by the enemy to such a degree that if captured they would face a firing squad without explanation. By their trench mates they were distrusted.' It goes without saying that they were hated by the enemy, which is why snipers somewhat fatalistically came to accept that in wartime the Geneva Convention applied to everyone except them. If they were captured, summary shooting was the best they could hope for. A Sergeant Powell of the Royal Welch Fusiliers saw first-hand what could happen to a captured sniper:

A lot of the sniping from houses had to be put up with, so had the shelling, for our guns had not enough ammunition to cope with them. But the circumstances in which some men had been shot led to the belief that there was a sniper behind us, so Stanway who had several snipers to his credit, took out a few men to make a search. While they were halted beside a large strawrick one of the men noticed some empty German cartridge cases at his feet. On thrusting their bayonets into the rick the party was rewarded with a yell and a German coming out headlong. Inside was a comfortable hide, having openings cleverly blocked with straw, and a week's supply of food. The sniper could come out at night for exercise and water. Only his carelessness with his used cartridges cost him his life, for he was finished there and then.

Obviously despised by the enemy, snipers were also disliked by their own side, not just because they were perceived as having luxurious privileges. Bill Howell, who was a sniper at Loos in the summer of 1915, reported that:

The men loathed us and the officers hated us. They could not order us out of their sector. The trouble was we could watch through a loop hole for hours, and when we were absolutely sure of a target we would fire. No other firing went on without an order and Gerry knew it was a sniper and he would let everything he had loose on that sector. Of course we hightailed it out, as fast as our legs would carry us, and poor old Tommy had to take it.

The sniper was at times not even subject to King's Regulations. Independent, resented, the sniper had to develop self-reliance, so that he was able to feed on others' dislike of him. He was a lone wolf.

Not every sniper revelled in his work. Privately many would concede that it was a dirty, murderous trade. Others were able to rationalise what they did to the point of mythologising it. They saw themselves almost as heroic warriors, whose battlefield duels were the earthbound equivalent of the courtly dog-fights of the early aviators. But many brave and cour-ageous soldiers, fierce and passionate in the heat of the battle, simply could not deal with sniping. Arthur Emprey was one such. Like many others he came alive in the heat of the fight. He thrilled to the notion of a bayonet charge and the one-on-one combat that it involved. But after killing two men as a sniper he knew he could not continue and had to ask to be transferred after only six weeks – 'To me it appeared quite all right to "get" a man in the heat of battle, but to lie for hours and days at a time, waiting for the enemy to expose himself – then to plug him, appeared to me a little underhanded.'

Frank Percy Crozier was described as a hard man, but he too found he had no stomach for sniping. Crozier had been a big-game hunter in Africa, but had no fondness for human targets: 'The game was dirty. I had to give it up. The cool calculating murder of defenceless men was diabolical.'

For some, sniping was simply a job. Nobby Clarke took very little pleasure in what he did, but he realised that it had to be done. The closest he came to dis-playing emotion was when one of his comrades was killed. Snipers by necessity were a close-knit group providing mutual support, and the death of one of them was felt keenly by all of his colleagues. 'When you lose one like we have in our sniper section you feel it for days,' said Clarke after the death of a comrade.

It is interesting to note that Clarke expresses only regret. There is no bitterness or rancour in his comments, especially not against any individual Germans. What anger he has he reserves for those in the High Command on both sides who were happy to order men to their deaths without facing any risk themselves, and in some cases without even coming to France.

For others the thrill of the kill was an occasion for triumphalism. Ion Llewellyn Idriess, known as Jack, was a trooper with the 5th Australian Light Horse fighting in the Middle East. After the war Idriess became a best-selling author with a series of books about his wartime exploits. In one of them, *Sniping*, he describes his feelings after hunting down and killing a Bedouin:

> Looking down on him like a great fallen hawk in the barley, I felt no remorse; only hot pride that in fair warfare I had taken the life of a strong man – hot pride that this man, older and stronger physically than I, this man reared from boyhood to regard warfare as the life of a man and splendid sport, this desert irregular, knowing every inch of the country, had fallen to a stranger from a peaceful land who had known only three years of war!

Idriess saw nothing underhand about sniping. Not for him the sort of moral equivocation that ultimately led the British Army to disband its sniper school.

Idriess saw honour in the job. He saw the sniper as a predator, out on the hunt.

The enthusiasm that Jack Idriess showed for his job is perhaps not surprising, since he learned his trade at the shoulder of one of the greatest and most lethal snipers to pick up a rifle.

While British and French troops were coming to terms with sniper fire on the Western Front, soldiers from Australia and New Zealand were being slaughtered by snipers on the far side of the Mediterranean. As conceived by the then First Lord of the Admiralty, Winston Churchill, the Allied landings at Gallipoli in April 1915 were a brilliantly inspired move on paper. The Gallipoli peninsula overlooked and controlled the Dardanelles strait, only 1 to 4 miles wide, and the only sea entry from the Mediterranean to the Sea of Marmara and the Turkish capital of Constantinople. If an Allied fleet could force the Dardanelles, then seizing Constantinople would open up the Black Sea to their Russian allies and relieve pressure on the Eastern Front as well as possibly providing a route into Germany via the Danube. In fact the campaign was a military disaster. The men of the Australian and New Zealand Army Corps – the ANZACs – were supposed to land on a shallow easy beach that would give them a foothold on the Gallipoli peninsula. Other landings by the French and British, which were intended as diversions, had indeed gone well. The Anzacs however were in light boats, which were towed by the powerful current to the wrong landing site, an anonymous beach that would for ever after be known as Anzac Cove in memory of the thousands of men who died there. The Gallipoli plans were an ill-kept secret, and the Turks had plenty of time to prepare their defences. When the Anzacs found themselves on a beach with steep mountains in front of them and the sea to their backs they were massacred by the enfilading Turkish machine-gun and sniper fire pouring down from the heights above.

Retreat was impossible. The men clung to a tiny strip of beachhead in order to maintain their foothold in Turkey. Although they suffered appalling losses themselves, they also killed thousands of Turks and their German advisers in a bitter campaign that lasted until the Allies finally withdrew in January 1916. The Turkish snipers, some of whom were reported to be women, were particularly deadly, but once they were dug in the Anzacs went about squaring the account. Billy Sing was one of those who plied his trade at Gallipoli, and he became so feared that the Turks put a price on his head and detailed snipers specifically to hunt him down.

William Edward Sing was born in Queensland in March 1886. His father was a Chinese immigrant, his mother an English nurse. As a young man Billy distinguished himself at the local rifle club, where he became a marksman with a Lee-Enfield, and when war broke out it was inevitable that he would join the Australian Imperial Force. He signed up in October 1914 along with a number of other Queenslanders including Jack Idriess. Both men joined the 5th Australian Light Horse and ended up at Gallipoli the following year. Sing became a sniper and Idriess was his first spotter. It is thanks to Jack Idriess's literary talents that we know about the legendary Billy Sing:

. . . a little chap, very dark, with a jet black moustache and a goatee beard. A picturesque looking mankiller. He is the crack shot of the Anzacs.

One of the hottest spots on the Anzac line was the trenches at Lone Pine, which were no more than 30 metres from the Turkish lines. These were to all intents and purposes a shooting gallery for the Turks. Corporal Gilbert of A Company, 13th Australian Light Horse, was one of those who remembered Lone Pine with little fondness:

My best mate and I used to go on the firing step together at Lone Pine. One morning moving into Lone Pine trenches one soldier just ahead of me turned to my mate and said 'Come on Dick, you and I will go on together this time.' One used the periscope to see what Johnny Turk was doing, the other was ready for any quick sniping at anything that moved. The rest of us waited in an old dug-out to take our turn. The next minute, bang. Dick got a bullet right through his head and he fell at our feet. He made no sound at all! He was still alive when the stretcher-bearers took him down to the beach to be put on a hospital ship for Malta. But he died there. We think an enemy sniper must have been just out in front using slight ground cover waiting for our relief guard to come in. I made sure I got that sniper later on.

Billy Sing always used a partner. With a spotter to constantly sweep the landscape, Sing could concentrate on sighting and shooting. He was unconsciously imitating the system that Hesketh-Pritchard was attempting to establish in France. He and Jack Idriess made a highly effective team. Sing's greatest asset, apart from his marksmanship, was his temperament. A man of remarkable patience, he could lie literally for hours waiting for the right shot. He and Idriess would leave their lines before dawn and head out into no man's land to pick the right place to shoot from. Sing would take up his position and wait. Idriess, who must also have been blessed with considerable patience, knew that Sing would not fire until he was certain of a kill. The spotter could wait until the unwitting Turk had exposed enough of himself to make sure the shot would be fatal. Once he heard Idriess shout 'Right', Sing squeezed the trigger and added another Turk to his tally.

The stories about Sing became the stuff of legend as his kill rate increased. One anecdote, which is probably a mixture of fact and fiction, relates that Sing was so impressive that his commanding officer General Birdwood asked to come out with him as his spotter. The story apparently comes from Birdwood, who claimed that when they were out together Sing fired at a Turk but apparently missed because he had not allowed for the wind. He fired again, shooting another Turk, but he refused to add that kill to his total because he hadn't hit his original target.

Even without that kill, Billy Sing personally accounted for more than 150 Turkish soldiers between May and September 1915. As the end of the year approached his total broke 200. He had become as notorious to the Turks as he was famous to the

Anzacs. On at least two occasions they sent marksmen out into no man's land with specific instructions to hunt down Sing and kill him. They came close. One Turkish marksman shot the telescope from the hands of Sing's new spotter, Tom Sheehan, wounding him in the face and hands. The bullet went on to strike Sing in the shoulder, but because it was almost spent it did not cause serious injury.

Those who were in the line with Sing say the incident plainly unnerved him and he was never quite the same man after it. Eventually he was posted to France when the AIF was evacuated from Gallipoli, but not before being awarded the Distinguished Conduct Medal for 'conspicuous gallantry as a sniper at Anzac Cove between May 1915 and September 1915'.

What distinguished Sing as a sniper was his attitude to his job. He killed without any apparent feeling. Some of the men whose lives he was undoubtedly saving through his work felt there was something distasteful about his craft. He eventually earned the nickname 'The Murderer' because of his apparently callous approach. The nickname seems to stem from one incident in which an old Turkish soldier had been trapped under a fallen trench support. The Australians thought the old man's struggles as he tried in vain to free himself were highly amusing. Sing simply took aim and shot him to 'put him out of his misery'. It did not matter to Billy Sing that the Turk was helpless and almost certainly going to die anyway; a target was a target.

In Turkey and France Billy Sing was shot twice, injured by shrapnel and gassed at least once. He was described as a shadow of his former self when he returned to Australia after the war. His personal and business life were little short of disastrous, and he ended up living with his sister in Brisbane, where he died of a heart attack in 1941.

In many ways Billy Sing conforms to the archetypal image of the sniper; the deadly assassin killing without compunction or remorse. Yet sniping and conscience are not mutually exclusive.

Alvin Collum York was America's most famous soldier of the First World War, a simple man from Pall Mall in Tennessee, barely three miles from the border with Kentucky. A wild man in his youth who had a fondness for drink, York had a religious epiphany at a revivalist meeting and swore off all of his former vices. Like Billy Sing, Alvin York was a keen marksman and developed his skills on the farm.

In our shooting matches at home we shot at a turkey's head. We tied the turkey behind a log, and every time it bobbed up its head we let fly with those old muzzle loaders of ours. We paid ten cents a shot and if we hit the turkey's head we got to keep the whole turkey.

We can assume that the York family ate turkey dinners perhaps more frequently than the other families in Pall Mall valley.

Sergeant Alvin York became world-famous on 8 October 1918 in the Argonne Forest in France when he single-handedly captured 132 German prisoners after his unit had taken heavy casualties and been pinned down. Although not defined by his regiment as a sniper, York merits inclusion on the strength of his marksmanship. He was close to the 'one shot, one kill' threshold. He kept a diary of his days in France. Although American officers discouraged the keeping of diaries or journals in case they contained any information that, if captured, could prove useful to the enemy, York held on to his thanks to a piece of moral equivocation. When he was asked if he had a diary he refused to either confirm or deny its existence, and the officer in charge didn't pursue the matter.

The diary contains York's account of the incident that made him famous:

I had no time nohow to do nothing but watch them-there German machine gunners and give them the best I had. Every time I seed a German I jes teched him off. At first I was shooting from a prone position; that is lying down; jes like we often shoot at the targets in the shooting matches in the mountains of Tennessee; and it was jes about the same distance. But the targets here were bigger. I jes couldn't miss a German's head or body at that distance. And I didn't. Besides, it weren't no time to miss nohow.

I knowed that in order to shoot me the Germans would have to get their heads up to see where I was lying. And I knowed that my only chance was to keep their heads down. And I done done it. I covered their positions and let fly every time I seed anything to shoot at. Every time a head come up I done knocked it down. Then they would sorter stop for a moment and then another head would come up and I would knock it down, too. I was giving them the best I had.

I was right out in the open and the machine guns were spitting fire and cutting up all around me something awful. But they didn't seem to be able to hit me . . . As soon as I was able I stood up and begun to shoot off-hand, which is my favorite position. I was still sharpshooting with that-there old army rifle. I used up several clips. The barrel was getting hot and my rifle ammunition was running low, or was where it was hard for me to get at it quickly. But I had to keep on shooting jes the same.

In the middle of the fight a German officer and five men done jumped out of a trench and charged me with fixed bayonets. They had about twenty-five yards to come and they were coming right smart. I only had about half a clip left in my rifle; but I had my pistol ready. I done flipped it out fast and teched them off, too.

I teched off the sixth man first; then the fifth; then the fourth; then the third; and so on. That's the way we shoot wild turkeys at home. You see we don't want the front ones to know that we're getting the back ones, and then they keep on coming until we get them all . . . Then I returned to the rifle, and kept right on after those machine guns. I knowed now that if I done kept my head and didn't run out of ammunition I had them. So I done hollered to them to come down and give up. I didn't want to kill any more'n I had to. I would tech a couple of them off and holler again. But I guess they couldn't understand my language, or else they couldn't hear me in the awful racket that was going on all around. Over twenty Germans were killed by this time.

. . . The next morning Captain Danforth sent me back with some stretcher bearers to see if there were any of our American boys that we had missed. But they were all dead. And there were a lot of German dead. We counted twenty-eight, which is just the number of shots I fired. And there were thirty-five machine guns and a whole mess of equipment and small arms.

His exploits in the Argonne made York a national hero. The incident was made into an Oscar-winning film, *Sergeant York*, with Gary Cooper playing the role of Alvin York. But throughout all of the attention he received he retained his natural modesty and always credited his faith and his maker for what happened in France. Indeed York used his own story as an example to others to embrace a Christian faith. He wrote in his diary:

So you can see here in this case of mine where God helped me out. I had been living for God and working in the church some time before I come to the army. So I am a witness to the fact that God did help me out of that hard battle; for the bushes were shot up all around me and I never got a scratch.

So you can see that God will be with you if you will only trust Him; and I say that He did save me. Now, He will save you if you will only trust Him.

After the war Alvin York used his new celebrity status to promote the cause of education for mountain children, and founded several educational institutes in his native Tennessee. He died on 2 September 1964 at the age of seventy-six. He is buried in Wolf's River cemetery, not far from the church where he was born again on 1 January 1915.

Temperament, as Hesketh-Pritchard had discovered, was perhaps the key element of a great sniper: the quality that could not be taught. To do the job to its deadly best it was necessary to be able to detach oneself from the killing shot. The sniper could not be tormented by doubt or haunted by nightmares. Some felt it was dirty work and had to abandon it, while those who excelled had to find coping mechanisms. Alvin York's religious conviction enabled him to reconcile what he was doing. Others found their own ways to deal with the job of being a sniper. Jack Idriess, for example, saw it as heroic combat to be revelled in, while his partner Billy Sing seems to have simply repressed his feelings.

Unlike other combat troops, the sniper is an individual decision maker. He is not bound by personal orders and therefore, as Hesketh-Pritchard was quick to realise, the individual psychology of the sniper mattered much more than in the ordinary soldier. This was certainly the case in the First World War, and the psychology of the individual gunman would go on to dominate sniping in the remaining years of the twentieth century and beyond.

12

A LESSON NOT LEARNED

The military authorities appear to have classified sniping as a tactic particular to trench warfare, and within years of the end of the First World War all of the combatants involved had disbanded their sniper schools and specialised sniping units, just as the sharpshooter regiments were stood down at the end of the American Civil War. No matter how effective they have been in battle, the minute hostilities cease civilised society appears to become embarrassed by those who hunt and kill without passion.

By 1921 the British Army ordered its stocks of the sniper-equipped SMLE Lee-Enfield rifles to be broken up and the telescopic sights sold as Army surplus. Germany too, under the Weimar Republic, had decided that there was no future in sniping. The German Army decided that its sniper rifles were to be used up and would not be replaced, nor would any spare parts be ordered to maintain them. Without the equipment there was no point in maintaining any sniper schools, and of the postwar world's major armies only the United States Marines, who had always prided themselves on their marksmanship, still had a few trained snipers.

Some countries persisted: Russia saw the value of snipers and retained their marksmen. The end of the war meant there were thousands of crack shots now out of uniform. Drifting back into civilian life all over Europe, many would soon take up their skills again in the Spanish Civil War. What better weapon for a politically motivated but poorly funded organisation than a sniper who, with a handful of bullets, can wreak havoc on the enemy? The myth of the sniper made him the perfect guerrilla weapon.

Snipers featured prominently on both sides in the Spanish conflict. The International Brigades would have included many men who had seen service on the Western Front and were able to reapply their skills with the same bloodily effective results. Jack Jones, who would go on to become one of the most powerful and influential figures in the British trade union movement, fought in the Civil War as a young man.

There were many casualties and I became one of them. Once more I had clambered up the hill with my comrades, taking cover where we could and firing at the enemy wherever he appeared. The bullets of the snipers whizzed over, grenades and shells were striking the ground, throwing up earth and dust and showering us with shrapnel. Suddenly my shoulder and right arm went numb. Blood gushed from my shoulder and I couldn't lift my rifle. I could do nothing but lie where I was. Near me a comrade had been killed and I could hear the cries of others, complaining of

200

their wounds. While I was lying there, to make things worse, a spray of shrapnel hit my right arm. The stretcher-bearers were doing their best but could hardly keep up with the number of casualties. As night fell I made my own way, crawling to the bottom of the hill. I was taken with other wounded men down the line to an emergency field hospital at Mora del Ebro where I was given an anti-tetanus injection. The place was like an abattoir; there was blood and the smell of blood everywhere.

There were others who could also attest to the effectiveness of snipers in the house-to-house fighting in the Spanish towns and cities during the Civil War. The British Labour MP Manny Shinwell was among a group of Parliamentarians who had been invited to Spain by the Republicans to monitor the progress of the war.

We walked along the miles of trenches which surrounded the city. At the end of the communicating trenches came the actual defence lines, dug within a few feet of the enemy's trenches. We could hear the conversation of the Fascist troops crouching down in their trench across the narrow street. Desultory firing continued everywhere, with snipers on both sides trying to pick off the enemy as he crossed exposed areas. We had little need to obey the orders to duck when we had to traverse the same areas. At night the Fascist artillery would open up, and what with the physical effects of the food and the expectation of a shell exploding in the bedroom I did not find my nights in Madrid particularly pleasant.

The Spanish Civil War almost immediately preceded the Second World War. By the late 1930s Germany had resumed sniper training at the insistence of Heinrich Himmler, who was a keen advocate of sniper tactics, although the Wehrmacht still did not have enough rifles. But when the Second World War began it was the only one of the initial combatants with a trained sniper force. Just as in France in 1914, the Germans went into combat with a theoretical tactical advantage. This time, such was the success of the German blitzkrieg, which rolled largely unopposed through Western Europe, that there was little need for German snipers to be deployed.

Ironically, in the early days of the war the army with the most trained snipers found itself being sniped at. In France the old tradition of the franc-tireur was revived by the Resistance movement. Francs-tireurs were civilian snipers who had their origins in a resistance organised against the German occupation of France in 1870. Germany had further experience of the francs-tireurs in the 1914 invasion of France and Belgium. They came to believe that these civilian sharpshooters had been trained and hidden in almost every village. In fact there were very few genuine francs-tireurs in the First World War, and the Germans tended to exaggerate their numbers in order to carry out reprisals to suppress the civilian population. By the Second World War there were undoubtedly more Resistance snipers, but their effectiveness was limited by lack of numbers and equipment.

Although they had been dismissive of the need for snipers after the First World War, the British Army reacted with commendable speed at the start of the Second World War. A new sniping school was set up, this time at Bisley, and the redoubtable Lovat Scouts were once more pressed into service. The crash course in field craft, observation, camouflage and shooting skills meant that snipers qualified for active service could be returned to the line in three weeks.

The early part of the war was not really suited to sniper combat. Snipers certainly had a role to play in delaying the oncoming German forces during the British evacuation at Dunkirk, but it was not until the Allied forces began to carry the war to Germany that the sniper really earned his place in combat. By the time of the D-Day landings both sides had large numbers of trained snipers, many of them already battle-seasoned in the Italian and Sicilian campaigns. The Americans had also opened their own sniper training school at Fort Bragg in 1942, and that meant they no longer had to rely simply on the Marine Corps.

Captain Clifford Shore was one of Britain's foremost authorities on sniping during the Second World War, and his memoirs have become almost a standard text for anyone interested in the subject. He was among the British troops who arrived at Normandy on D-Day, but because there was not enough transport to get everyone off the beaches he and his comrades had to spend the night in their landing ship. The craft, which was large enough to carry tanks, pulled back to about a mile off shore, where its size began to attract the attention of German dive-bombers. It was strafed and hit more than 100 times, two men were killed, and the ship was burning below decks. It was, as Shore recalls, an awful experience:

I can never forget that night; the sense of oppression; the choking in the throat from the smoke generators which had been lighted on deck to 'blanket' the craft from the Hun planes; the demon wailing of ships' sirens; the headlong rush for the sparse cover; the sudden roar of diving planes; the scream of falling bombs; the convulsive lurch of the ship as the bombs tore the water and the sides of the craft; the tearing of steel and the moans of the wounded with the pearls of shock sweat on their forehead. It was the longest night of my life!

Throughout Shore's recollection of the war there is an almost stereotypical stoicism about his exploits, almost as if he is unwilling to admit how terrifying the whole experience must have been. Even when they finally made it to the beach on D-Day plus 1, Shore remains coolly unemotional:

In Normandy my experiences of sniping were more of the receiving rather than the giving variety. Shortly after landing I saw an officer leave the Assembly area in a Jeep, smiling broadly, driven by a harassed faced driver. A few minutes later the Jeep returned with the officer dead, a neat hole being drilled in the centre of his forehead. I must admit that it gave me and the others a shock. Apparently it was not going to be the 'piece of cake' we had been led to believe at the numerous briefings to which we had listened with great interest and growing enthusiasm. I noticed that when the men brewed the inevitable 'char' they kept very close to the trucks and showed very little inclination to stand upright.

D-Day was not 'a piece of cake', as Shore puts it, for anyone involved. Outnumbered and taken by surprise, the Germans poured fire down on to the Normandy beaches, knowing they had little chance of turning back the invasion but hoping to delay the Allied timetable. It was the ideal job for the well-trained German sniper.

At Omaha Beach, one of two locations where the Americans were coming ashore, 400 yards of open ground separated their landing craft from the relative security of the sea wall, the equivalent of almost four football pitches. German snipers tried as best they could to keep the Americans pinned down within this generous killing ground. This involved deliberately targeting corpsmen and medical staff. These brave men could not have been more conspicuous as they darted around the beach tending to the thousands of wounded. The red crosses on their armbands and on their helmets made perfect targets. The Germans shot at the medics as they attempted to drag wounded men from the beach to the shelter of the sea wall.

One of the worst-hit sectors of Omaha Beach was the area designated Easy Red where the US First

Infantry Division came ashore. The Americans lost almost all of their officers and about half of their non-commissioned officers. War correspondents who followed the troops ashore reported that there was scarcely a piece of beach without a dead or wounded man on it. They lay so thick on the ground that it was almost impossible to take a step. They were soon hours behind a schedule that called for them to be off the beaches in a matter of moments.

V Corps, which landed at Omaha Beach, saw 2,400 of its 34,000 men dead, wounded or missing on D-Day, a casualty rate of 7 per cent in a single day. But as Stephen Ambrose points out, those figures do not tell the whole story. The bulk of the men who landed on Omaha Beach arrived in the late morning or afternoon, but most of those casualties came in the first hour of fighting.

No sooner had the Americans flushed the Germans out of one location than they would use the warren of communications tunnels linking their gun emplacements to take up fresh positions. Things were no better once the invasion force got off the beach and made their way inland towards Colleville. Again, in the ghastly phrase much loved by those who ply the trade, the German snipers did great execution. Sergeant John Ellery was one of the more fortunate:

I was about to climb through a break in the hedgerow when my ID bracelet on my right wrist got hung up on a rather sturdy piece of bush. I slid back down and broke off the branch to get loose. Meanwhile a fellow from another company decided to pass me by and go on over. As his head cleared the top of the hedgerow, he took a round right in the face, and fell back on top of me, dead.

Ellery found the sniper and avenged the man who had taken the bullet that should have been meant for him. He killed the sniper with one clean shot; the only shot he fired on D-Day.

At Sword Beach, where the British Third Infantry Division was landing, the German snipers had a plum target. Brigadier Lord Lovat, the commanding officer of the British Commando Regiment, defied all attempts by German snipers to shoot him. The death of Lovat, whose family name was adopted by the Lovat Scouts, would have been a major blow to the spirit of the invading Allied troops, but he continued to swagger around the beach in a morale-boosting show of defiance. As shot and shell whizzed around him he only encouraged his piper to play louder, assured his men it was no more dangerous than an exercise, and strode unarmed towards the objective. His piper, Bill Millin, was playing in defiance of a general order from High Command, who believed that the pipers would attract sniper fire. The then 21-year-old Millin recalls the moment:

I was the only one with a kilt, a set of bagpipes and a knife, and I wasn't armed. My most traumatic experience was jumping into the cold waters with a kilt on. As I hit the water I began playing 'Highland Laddie'. Lord Lovat turned round and gave me a smile.

Bill Millin was decorated for his courage. His bagpipes, which were eventually silenced by enemy shrapnel some days later, are now in the National War Museum of Scotland at Edinburgh Castle. He was able to render one final service to his beloved commander when he played the lament at Lovat's funeral in 1995.

Once the Allies made their way off the beaches they found themselves in one of the most beautiful parts of France. The Bocage region of Normandy is a timeless relic of rural tranquillity. Very little seems to have changed since. The area is characterised by its distinctive hedgerows, huge earthen banks surmounted by thick trees and bushes. These are man-made redoubts, perfectly described one hundred years earlier by the French writer Balzac:

The peasants from time immemorial have raised a bank of earth about each field, forming a flat-topped ridge, two metres in height, with beeches, oaks, and chestnut trees growing upon the summit. The ridge or mound, planted in this wise, is called a hedge; and as the long branches of the trees which grow upon it almost always project across the road, they make a great arbour overhead. The roads themselves, shut in by clay banks in this melancholy way, are not unlike the moats of fortresses.

Although they are now in decline thanks to modern roads and growing urban development, the hedgerows of Normandy can still be spectacularly beautiful.

In June 1944 when they were still relatively unspoiled, not only were they beautiful, they were extraordinarily dangerous. As Balzac says, the roads running between these hedgerows were like moats or dry canals, and there was no option for Allied troops but to take these roads. They would eventually find a way of breaching the thick banks of the hedgerows, but to begin with all they could do, if they wanted to maintain the momentum of their advance, was use the roads.

The German snipers were expecting them. Units comprising two or three snipers were left in hiding to cover the retreat as a determined rearguard to delay the advancing troops while the German forces regrouped elsewhere. The Americans were good marksmen but less skilled in field craft; the Germans were masters of both. They would wrap their rifle barrels in burlap to prevent their position being betrayed by a telltale glint of sun on metal, they wore camouflaged uniforms, and constructed elaborate hides in the trees that they then covered with camouflage netting. They were extremely difficult to detect and could pick off their targets with impunity. A German sniper, or in some cases a two-man team, could live in the hedgerows with enough food to last several days. An individual soldier walking down the road alone would be shot and killed instantly. The appearance of a group of soldiers would lead to artillery, which had often been pre-sighted, being called in by the snipers, in a clear development of the tactics laid down by Hesketh-Pritchard during the First World War, when British snipers would be used as observers for artillery barrages during major operations. The sniper was now integrated into a chain of combat and was becoming much more than just his rifle and bullet.

A single sniper could delay the advance for hours or more. It was difficult for Allied commanders to keep their troops moving in the face of sniper fire. The natural reaction is to halt, even though that only presents more targets. NCOs and officers tried to instil in their men the instinct to drive forward in the face of sniper fire, believing that a platoon that freezes when one man is killed often leads to many more being picked off where they stand.

Having lost so many officers in the Normandy landings, the Allies did their best to make sure that they kept the ones who had survived. In an echo of the tactics employed by Union commanders in the American Civil War, officers were instructed to cover up their insignia and badges of rank so they could not be identified. They were also encouraged to walk in the middle of the column rather than at its head, in the hope of concealing their identity from snipers.

Still the system was not infallible. Clifford Shore recalls that he had ordered his officers and NCOs to cover their chevrons when they landed in Normandy. He later found that his orders had been countermanded by his colonel, who insisted that badges of rank be displayed; if they were to be killed, he insisted, then they would be killed as officers. Just as they had in the First World War, German snipers again had a formidable strike rate when it came to identifying and targeting British officers. On the Western Front they had shot the ones with thin legs. In Normandy they aimed for those with moustaches, since only officers were allowed facial hair!

The life expectancy of a German sniper, however, was not long. If they were spotted then they could expect to be blown out of the hedgerows by machine-gun fire. If they were able to escape undetected then they had to try desperately not to be taken prisoner. They wore no badges of rank or regimental insignia and travelled light deliberately in the hope that if they were captured they could pass themselves off as infantrymen cut off from their units.

Some Germans were more fortunate than others. For his book *D-Day* the American historian Stephen E. Ambrose interviewed a number of D-Day survivors, including Lieutenant Jay Mehaffey of the US Rangers. Mehaffey and his men had just lost a comrade in Vierville, cut down by a German sniper as he passed a gap in the hedgerow. Just then another Ranger came along with eight German prisoners. Mehaffey lined the Germans up across the gap where his man had been shot; using them for cover he then filed the remainder of his detail past the gap behind the German prisoners. Once his own men were across safely Mehaffey, who had no time to take prisoners, waved the Germans off down the road and told them to give themselves up to the first American unit they found.

The voice of the American infantryman in the Second World War was that of Ernie Pyle, one of the great war correspondents. His dispatches from the front humanised the war for millions of people the world over. His column appeared in more than 400 newspapers worldwide, and he was so well regarded that his war exploits were turned into a film, *The Story of G.I. Joe*. Pyle summed up the views of many towards snipers in one of his reports a few days after D-Day:

Sniping, as far as I know, is recognised as a legitimate means of warfare. And yet there is something sneaking about it that outrages the American sense of fairness. I had never sensed this before we landed in France and began pushing the Germans back. We had had snipers before – in Bizerte and Cassino and lots of other places, but always on a small scale. There in Normandy the Germans went in for sniping in a wholesale manner. There were snipers everywhere; in trees, in buildings, in piles of wreckage, in the grass. But mainly they were in the high, bushy hedgerows that form the fences of all the Norman fields and line every roadside and lane.

It was perfect sniping country. A man could hide himself in the thick fence-row shrubbery with several days' rations and it was like hunting a needle in a haystack to find him. Every mile we advanced there were dozens of snipers left behind us. They picked off our soldiers one by one as they walked down the roads or across the fields. It wasn't safe to move into a new bivouac area until the snipers had been cleaned out. The first bivouac I moved into had shots ringing through it for a full day before all the hidden gunmen were rounded up. It gave me the same spooky feeling that I got on moving into a place I suspected of being sown with mines.

In past campaigns our soldiers would talk about the occasional snipers with contempt and disgust. But in France sniping became more important, and taking precautions against it was something we had to learn and learn fast. One officer friend of mine said 'Individual soldiers have become sniper-wise before, but now we're sniper conscious as whole units.'

Snipers killed as many Americans as they could, and when their food and ammunition ran out they surrendered. Our men felt that wasn't quite ethical. The average American soldier had little feeling against the average German soldier who fought an open fight and lost. But his feelings about the sneaking snipers can't be put into print. He was learning how to kill the snipers before the time came for them to surrender.

There is no doubt that there were snipers on both sides who were shot when they were taken prisoner, just as there is no doubt that there were other prisoners of war on both sides who were shot summarily rather than accorded the rights of the Geneva Convention. The evidence for the execution of snipers and other prisoners is largely anecdotal: no charges were brought on the Allied side. General Omar Bradley, the commander of the American forces in Europe, seems to have tacitly condoned the practice. He let it be known that he believed snipers were not playing the game and he had no objection if they were treated harshly.

Harry Furness was a British sniper in Normandy. He says both sides were well aware of the practicalities of the situation.

All snipers if captured were shot on the spot without ceremony as snipers were hated by all fighting troops; they could accept the machine gun fire, mortar and shell splinters flying around them . . . but they hated the thought of a sniper taking deliberate aim to kill by singling them out.

Furness offers an interesting psychological insight. By his logic it would be perfectly acceptable to shell a tank and the crew inside, but completely unacceptable to disable that tank by singling out the commander and shooting him alone. What he hints at is the common conviction that war in the abstract – against a building or machine for instance – is bearable while war that focuses on the individual combatant is somehow abhorrent, inhumane. This perhaps is what lies fundamentally at the heart of the soldier's objection to the sniper. As far as his target is concerned the sniper has made war personal. Ironically, many snipers seem to be able to deal with the psychological burden of their craft by being resolutely impersonal, at times almost robotic, in carrying out their duty.

The British sniper Arthur Hare, for example, in his description of his victims is quick to dehumanise them. A corporal sniper, Hare often found himself as an advance scout for artillery units. On one occasion in Holland, he and three colleagues had been ordered to a farmhouse to provide cover for a British attempt to cross a river. They waited quietly and unobserved from early in the morning until the British started shelling German positions just after two in the afternoon. As the Germans ran for cover, Hare and his fellow snipers began their work. Inside ten minutes they had accounted for some 200 Germans. When the battle was over they went across to look at the bodies of their targets.

Their faces were waxy, eyes staring, some with triumphant grins, others frozen for all time . . . fear, anger, despair, showed plainly on their features. Some old, some young, all battle weary and now hardly even human.

Arthur Hare was awarded the Military Medal for his part in the action.

Captain Clifford Shore had no doubt about the value of the sniper in battle. Like Ion Idriess in the deserts of the First World War, he saw it as a noble and honourable calling, not the action of an ignoble or cowardly soldier:

There are still many unenlightened people who think that sniping is 'dirty', 'horrible', 'unfair'. I maintain that it is the highest, cleanest game in war. It is personal and individualistic; a game of great skill and courage, of patience and forbearance. And if one loses to a skilful opponent one should never be conscious of it. Death-night to the sniper's quarry should always be of a tropical, sudden intensity.

The slow progress of the Allies through the hedgerows of Normandy was almost a sprint compared to their counterparts in the Pacific theatre. For the Japanese, surrender was dishonourable, unthinkable; every metre of land they had gained in 1940–42 would have to be defended to the last breath. Japanese snipers dedicated themselves to holding back the Allies at every turn. With complete disregard for their personal safety, they chose tactics that at times verged on suicidal.

Although the German snipers in Normandy constructed hides in the hedgerows, they rarely chose to site themselves in the tops of trees, since a sniper discovered in the treetops has no means of escape. The particular revulsion that ordinary troops had for snipers ensured that once detected they were almost certain to be shot out of the tree. The Japanese had no reservations about placing their marksmen aloft.

One factor in the Pacific theatre that favoured snipers was the density of the tropical forests, near impenetrable in places. Japanese snipers, issued with three-pronged climbing spurs so that they could climb the smooth trunks of palm trees to take cover in the dense crown of fronds at the top, built elaborate hides. Sometimes they even incorporated seats into their design so that they could stay there longer, roping themselves to the tree to stop them falling if they were wounded or fell asleep. Once concealed, they would wait for Allied troops to pass below them. They became adept at using bird or animal calls to disguise any noise they might make as they changed position. With green uniforms and hands and face also painted green, the Japanese snipers were extremely difficult to discern against the jungle canopy. The use of camouflage netting made them almost invisible even at short distances, unless of course they moved.

War correspondent Richard Tregaskis recalls one

close encounter on Guadalcanal with a Japanese sniper. He was sitting on a ridge watching Japanese Zeros and American Grumman fighter planes in an aerial dogfight trying to get some idea of what was going on.

Suddenly I saw the foliage move in a tree across the valley. I looked again and was astonished to see the figure of a man in the crotch of the tree. He seemed to be moving his arms and upper body. I was so amazed at seeing him so clearly that I might have sat there and reflected on the matter if my reflexes had not been functioning – which they fortunately were. I flopped flat on the ground just as I heard the sniper's gun go off and the bullet whirred over my head. I knew then that his movements had been the raising of his gun.

His fellow correspondent John G. Dowling of the *Washington Post* said that there was no option but to feel like a fool when there were snipers around. There was, as he pointed out, always the chance that you could be a target and, unlike Tregaskis, you might not be fortunate enough to see it coming. 'It is beneath your dignity to fall on the ground and crawl, yet standing there you feel like a balloon in a shooting gallery.'

Prudence won out over dignity every time as the correspondents and the troops on whom they were reporting inched through the jungles of the Pacific islands. One of their greatest challenges was fatigue, as Captain John Monks of the US Marine Corps recalled. The heat, long marches and the sheer effort of concentration were a dangerous combination.

> . . . tired men become dead men, for tired men grow careless. And they knew it. They strained to pierce the blank scrim of the jungle night, and listened intently to the million jungle noises – ground noises particularly, for they must learn to distinguish the sound of a lizard or a rat or a toad, or any one of the hundred little ground animals in the jungle, from the even belly-slide of a crawling Japanese jungle fighter.

The tree-bound Japanese were in a vulnerable position, and this dictated their tactics. If they shot at a group heading towards them there was every chance that they could be located and fire returned. A frequent tactic was to wait until the unit had passed their tree before opening fire from the rear. The preferred targets were officers and NCOs, whom they could identify from their insignia. Sniping from behind increased confusion among their victims and made their own position that much harder to detect. Additionally the Japanese snipers were assisted by their use of small-calibre bullets, no less deadly but quieter and producing less smoke than conventional ammunition.

In an attempt to combat the problem the United States War Department issued a number of intelligence bulletins to their soldiers fighting in the Solomon Islands in 1942. Bulletin number three, issued in November 1942, contains some concrete advice on how best to deal with the Japanese snipers. The information came from a Marine officer with first-hand experience:

> They shot at us from the tops of coconut trees, slit trenches, garden hedgerows, from under buildings, from under fallen palm leaves . . . One sniper, shot down from a tree, had coconuts strung around his neck to help conceal him. Another in a palm tree had protected himself with an armour plate. Our Thompsons and B.A.R.s [Browning antomatic rifles] proved to be excellent weapons for dealing with snipers hidden in trees.

The American method of combating Japanese snipers could be brutally effective. Once a sniper had been spotted in a tree top, or a tree top had been spotted which looked as if it might contain a sniper, troops simply poured machine-gun fire into the suspect branches until the sniper was killed or the tree made useless for anyone else to conceal themselves. This was the origin of the tactic used by the Americans in Vietnam known as 'reconnaissance by fire': shoot at a bush to see if it shoots back.

By 1944 the Americans were adopting a brutally excessive approach to the Japanese snipers, grimly deciding that the best anti-sniper weapon was a howitzer. They simply destroyed the snipers and the trees along with them in a display of mass firepower. This proved successful, but only after the sniper had been detected. He was still often allowed the first shot.

The British in Burma also used mass firepower but generally preferred organised counter-sniping expeditions. They used conventional tactics of sending out two-man teams of sniper and spotter to patiently root out and eliminate the threat from the Japanese snipers. On one occasion two units combined to put twenty-four sniper teams into the field, accounting for 296 Japanese casualties with the loss of only two of their own men.

The war in the Pacific was ended by the least precise, most excessive weapon ever deployed – the atomic bomb dropped on Hiroshima on 6 August 1945. The Allied and American justification for the bomb was that it hastened the war's end and saved American lives. It was a weapon that contrasted utterly with the spirit and practice of warfare that the individual sniper waged. One American life, however, was not saved by the bomb dropped from the *Enola Gay*. After the war in Europe had ended, Ernie Pyle was sent to cover the war in the Pacific. He was reluctant to go; friends said he had a premonition of death. But Pyle, the constant champion of the GI, could not find it in his heart to abandon the many young Americans who, like him, had no wish to be far from home, battling in the Pacific.

On 18 April 1945 Pyle, still fairly new to the Pacific Theatre, stepped ashore on the island of Ie Shima just to the west of Okinawa. He was with a small group of GIs and he had barely set foot on the beach when he was killed. Ernie Pyle, the man who felt that sniping outraged the American sense of fairness, had been shot by a sniper.

When news of Pyle's death reached him, General Omar Bradley said he had known 'no finer man, no finer soldier' than Ernie Pyle. The legendary correspondent would probably have been more moved by the tribute paid by the soldiers who were with him when he died. The man who had seen so many dog tags on so many makeshift graves was commemorated with a simple plaque.

At this spot, the 77th Infantry Division lost a buddy, Ernie Pyle, 18 April 1945.

13

ENOUGH TO BURY
THE DEAD

One of the great battlefield maxims is know your enemy. The good sniper makes this almost impossible. The sniper's greatest friends are invisibility and anonymity. He should come and go with only the body of an enemy soldier to mark his passing. His versatility and mobility make it impossible to truly predict his movements or where he might strike next, bringing confusion and frequently panic to the enemy.

However, sometimes the sniper, who can be said to embody the lone figure fighting in defence of the homeland, is a valuable symbol, a propaganda asset. In 1942 a sniper was cloaked not in anonymity but in mythology. That year the Soviet Union found hundreds of miles of its territory occupied by German troops and the symbolically significant city of Stalingrad under siege. The lightning offensive of Operation Barbarossa had seen German troops pour across the borders almost unopposed since the summer of 1941. As the twenty-fifth anniversary of the October Revolution approached, the Soviets were in desperate need of something to boost morale. They chose the image of the sniper.

Suddenly these men and women were lauded as heroes of the Soviet Union. They were encouraged to kill as many Germans as possible and their exploits were reported in glowing terms in propaganda news-sheets and Party broadcasts. They were even decorated according to how many Germans they had managed to dispatch.

The Soviet Union knew all too well about the impact that snipers could have. Soviet observers of the Spanish Civil War had returned with glowing reports about the snipers in the International Brigades. But they had also seen evidence of the sniper's force closer to home. The lesson learned would not be easily forgotten.

The non-aggression pact between Hitler and Stalin in 1939, though little more than a sham, had allowed both sides to pursue expansionist policies in whatever areas they chose so long as they did not actually attack each other. Hitler turned his eyes towards Poland while Stalin had Finland in his sights. In November 1939 the Soviet Union engineered a border dispute and demanded that Finnish troops withdraw. To no one's surprise the Finns refused, giving Stalin the pretext for abandoning neighbourly neutrality and planning an invasion.

The Red Army in 1939 was still recovering from a huge purge of senior and mid-ranking officers during the Terror of the 1930s. Experienced commanders needed for the invasion were scarce. Lack of planning meant ordinary soldiers were lamentably underprepared. Many of the troops who would take part in the invasion were stationed in the Ukraine, which is cold in November but not to the arctic extremes that would be found in Finland. The Red Army was deployed in a hostile, unfamiliar terrain with next to no intelligence about its enemy.

Still, the Soviets had a significant numerical advantage. The entire Finnish Army, spread throughout the country, numbered around 600,000 men. The Soviets began the campaign with just under that figure but quickly brought in reinforcements, and within months they had some 900,000 men in the field. The Finns were also short of heavy artillery and ammunition: the average Finnish division had 3,000 fewer men and about one-third less artillery than its Soviet counterpart.

It was obvious that in direct conflict the Finns would have been overrun, but they had no intention of fighting face to face. Instead they deployed an army of highly motivated individuals who were determined to keep their country a sovereign state. The Finnish sniper came to embody the essence of that highly motivated army.

Like the American colonists or the German Jägers, many Finns hunted for a living. They knew the terrain, they understood the conditions, and even without telescopic sights – they employed conventional open-sighted Mosin-Nagant rifles – they wreaked havoc among the invading Soviet troops. The Mosin-Nagant is the workhorse of the sniper rifle family – it is Russian-made – and the Soviets would again face its deadly reliability in their more modern excursions into Afghanistan.

In the so-called Winter War, nowhere was the carnage more complete than at Suomussalmi in the far north of Finland. The Finns' 9th Rifle Division was under attack from two Soviet divisions – the 44th Rifle Division from the east and the 163rd Rifle Division from the north. Atrocious weather conditions prevented the Russians from moving up in force. As it made its way from Kiev the 44th Division was strung out along the road like a ribbon, almost in single file. Finnish snipers and ski troops could attack at will, picking their targets as and when they liked. Impossible to spot in their camouflaged white uniforms, they could glide off into the snow-covered landscape ready to set another ambush. A Finnish counter-attack drove the 44th back, and as they retreated in disarray they were slaughtered by the harrying gunmen. The 163rd met a similar fate when it tried to help by launching a pincer movement. Casualties were so heavy that the two divisions effectively ceased to exist after Suomussalmi.

From similar encounters two Finns passed into the annals of sniping legend. Both were unassuming farmers who, good citizen soldiers both, simply took up their rifles and went to hunt the Soviet invaders.

Simo Hayha left his farm and joined the 34th Infantry Regiment. He applied himself to his new task with a will, and in a nine-month period he was credited with 505 kills, the most by any sniper in any combat in history. When the war ended he went back to his farm, where he enjoyed a long and contented life until he died in April 2002.

Almost as effective as Hayha was Suko Kolkka. Also equipped with an open-sighted Mosin-Nagant, Kolkka treated sniping as a sport. He was especially fond of sneaking behind the Soviet lines to create panic and fear among the invaders' ranks. Kolkka is credited with just over 400 kills in 105 days of combat as a sniper. He was also said to be fond of using a machine gun, with which he killed a further 200. Kolkka became such a threat to the Soviets that snipers were sent to hunt him down. He outstalked, outhunted and outshot them all, one with a clean shot at 600 yards after the two men had stalked each other over frozen ground for days.

Like the Boers in South Africa, the Finns fought the Soviets to a standstill; they were outmanned and outgunned but they could not be outfought. A face-saving truce was signed in March of 1940, and although Finland ceded a large part of the territory that the Red Army had occupied, Finland remained a free nation.

The Soviet Foreign Minister Molotov claimed that 48,745 Russian soldiers had been killed, with a further 158,000 wounded in the conflict. Other less partial sources suggest that the Soviets lost 333,084 men with 65,384 dead, 186,584 wounded and the rest either missing, sick or incapacitated through frostbite. The Finns lost just over 25,000 men. At the end of the fighting Stalin had gained perhaps 20,000 square miles of Finnish territory for his Soviet empire. As one of his generals bitterly put it: 'Just enough to bury our dead.'

This Pyrrhic victory for Stalin would have another, much greater, cost in the years to come. Hitler had been watching the outcome of the Winter War with interest, and the Soviet military performance encouraged him to believe that the country would be vulnerable to invasion. Hence in June 1941 he launched Operation Barbarossa. Millions of German troops swept almost unopposed through the country, with the only serious obstacle according to some veterans being the dust from their own tanks during the hot dry summer.

The Red Army eventually regrouped after a series of crippling reverses, but before it was able to do that the Army had good reason to be grateful to its snipers. Soviet marksmen and women successfully fought a rearguard campaign that harassed and held up the German invaders as well as inflicting significant casualties.

The mark of a Soviet sniper was the Voroshiloff Sharpshooter badge. By 1938, just before the start of the Winter War, the Soviets claimed to have some 6 million soldiers who had qualified as snipers. This was a statistic viewed with some suspicion by other snipers, such as Britain's Captain Clifford Shore. He suggested these were not real snipers and that the Voroshiloff badge was merely a marksmanship quali-fication, something he described as 'sharpshooting in its lowest form'.

Despite Shore's rather sniffy dismissal it is still an impressive number of shooters – even if they are only qualified to a basic level – to be able to put into the field. The biggest problem facing the Red Army was the inexperience of its snipers. It was not uncommon to see a veteran with a posse of protégés trailing along behind hoping to live long enough to learn from his wisdom and field craft.

The large numbers of snipers who could be put into the field at very short notice very quickly made their mark. The best were cunning, ingenious and constantly inventive. In September 1941 one German

Panzer unit had been sniped at continually for a five-day period with heavy losses. Finally one morning a German spotter saw what he took to be a wisp of smoke coming from a burned-out Russian tank. In fact it turned out to be the breath of the sniper fogging in the morning chill. He had been hiding in the knocked-out T-34 for a week, living off the dead crew's rations and killing Germans at will.

The Soviets took to the trees and proved savagely effective. Just as the Germans did in Normandy, they had come up with their own variation on the ghillie suit. Occasionally a fine net could be added under the hood to cover the face and make the sniper even harder to spot. An all-white variation was developed for fighting in snow – a legacy of the bitter lesson they had learned trying to find camouflaged Finnish snipers. Ultimately the Soviets became so skilful at camouflaging themselves that they were rendered almost invisible as they waited for the German invaders to come within range.

Otto Bense, a Wehrmacht sergeant, recalled the experience of being sniped at by the Soviets:

They were placed in the trees and they were camouflaged so we never knew if they were there or not. Their weapons had telescopic sights. They seemed to be all over the place, where you wouldn't imagine them to be. Once I said to one of my colleagues 'Be careful' then crack, ping, and he was dead. It happened in a moment.

Just like the Germans on the Western Front, the Soviet snipers had an uncanny knack of picking out key targets. They would concentrate on officers or radio operators or artillery spotters so that their shots would have maximum effect. Sergeant Bense knew that he was a potential target:

> Pretty well our only defence was to remove all our badges so that we couldn't be recognised as officers from a distance. The SS did the same. They had a special braid and they removed it. The SS were extra-special targets. The Russians would take prisoners when it was just the Wehrmacht, but if they got hold of SS men it was different. They killed them all, shot them out of hand.

The Germans had little effective response to the Soviet snipers. 'We were in the woods near Minsk,' Bense recalled. 'We saw no one, no one, and we had losses and more losses. Suddenly a man fell out of the tree in front of me, dead. He was in a camouflage outfit in a leaf pattern, and he had been seen by one of our men and shot down. He fell at my feet.'

The sniper had been directly above Bense, and had another German not spotted the shooter, Bense might have known nothing about it until he became his next victim. In one period of just a few hours in September 1941, the German 465th infantry lost 75 dead and 25 wounded to the tree snipers. By the time the Germans had rallied the Russians had gone, melting away to fight another day.

The Soviet sniper could strike anywhere, as Guy Sajer, who was posted to the crack Grossdeutschland Division, discovered. He and his comrades had been marching for three hours through the snow when they decided to take shelter in a building they could see in the distance.

It didn't seem a bad idea. We had regrouped and a young fellow covered with freckles . . . was joking with his friend. We were making our way towards the hut when a violent burst of sound struck my ears. At the same moment, I saw, to the right of the hut, a light puff of white smoke. Utterly astounded I looked around at my companions. The Feldwebel had thrown himself down on the ground like a goalie onto a ball, and was loading his automatic. The fellow with the freckles was staggering towards me with enormous eyes and a curious stupefied expression. When he was about six feet from me, he fell to his knees. His mouth opened as if he wanted to shout, but no sound came, and he toppled over backward.

Eventually Sajer's unit returned fire and the sniper was wounded and captured. Their attacker turned out not to be a Soviet soldier but a partisan sniper, who had been taking part in the guerrilla campaign against the Germans. As they were deciding what to do with him a German lieutenant arrived on the scene. 'He shouted an order to the two soldiers who were with him. They walked over to the unfortunate man lying on the snow and two shots rang out.'

Just as the British had done on the Western Front, the Germans felt there was something inherently unfair about the Soviet snipers. There was a consistency of reaction here among all combatants. Gerhard Munch fought in Army Group B, which suffered dreadful losses at the hands of Soviet snipers at Stalingrad.

The Russian sniper who worked in our sector is again and again celebrated as a major hero. I found it inwardly revolting. I always compared it with sitting in a raised hide and shooting deer – that's got nothing to do with soldiership in my personal opinion.

The Soviet marksmen and women were well equipped for the different conditions in which they found themselves fighting. As in other armies, they were singled out for preferential treatment. They were the best-supplied, best-provisioned and best-clothed men and women in the Red Army, which is one reason why there was never a shortage of applicants to sniper school. Not only was it a possible passport to enduring fame, it also meant you would seldom go hungry – a major consideration when half the country was starving.

The Red Army used the Moisin-Nagant rifle. A telescopic sight increased the accuracy of the standard rifle to about 800 metres, allowing one sniper to command a huge killing ground on the flat wastes of the steppes. When supplies of these original rifles dried up, the Russians cannibalised parts from shattered rifles to make cruder versions. Although not as effective they were still deadly at 400 metres. They proved ideal for the urban street fighting that was to come.

Once the war moved from the plains into the cities,

the snipers had everything they needed to practise their black arts. Bombed-out buildings and piles of tangled steel gave them the cover where they could lie for hours waiting to make the right shot. Since there was little food or water inside the ruined cities by the time they were occupied by the Wehrmacht, German troops were obliged regularly to break cover for supply runs. One particularly sadistic tactic at which the Soviet snipers became adept was to shoot soldiers who had been sent for water. They would not shoot as the unfortunate volunteer was going to get the water, but would wait for him to start his return journey. His death added to the mental torture of his thirsty comrades.

The heroic stand by Soviet troops and civilians in cities such as Leningrad and especially Stalingrad enabled the Russians to turn the tide of the war on the Eastern Front. These were almost like medieval city sieges. Stalin had told the people of the city named after him that there was no land beyond the Volga; this was where the invader would be stopped. The people of Leningrad were similarly determined; every metre of ground would have to be fought for. Bitter battles, raging for days, were fought over a single house in a single street, then refought for the house next door. By virtue of circumstance and political design the Soviet sniper was poised to become a cult hero.

A competition was announced across the entire Soviet Army to see who could kill the largest number of Germans. Anyone with forty kills to his or her credit was entitled to be known by the title 'noble sniper' as well as receiving a special medal for bravery.

The winner in the competition was a sniper known only as 'Zikan', who was responsible for 224 kills by 20 November 1942.

The Red Army employed women in all ranks. Many of those entitled to call themselves 'noble sniper' were women. Roza Shanina, for example, who looked the very epitome of solid Soviet womanhood, was a Senior Sergeant credited with 54 kills. She wore her Order of Glory with great pride.

The Red Army took advantage of the fact that women were able to kill just as ably and every bit as effectively as their male counterparts. Nina Lobkoskaya was a sniper with 89 kills. She led a company of female snipers from Russia to Berlin itself, and together they accounted for more than 3,000 Germans. She had originally volunteered to go to the Front as a First Aider or a wireless operator. As part of her training she was required to learn to shoot and quickly gained her Voroshiloff badge. After taking part in a shooting contest with several hundred other women she and a friend came top and were sent to sniper school.

'As soon as I started studying I realised that the mere desire to fight the enemy was not enough,' she recalled. 'You had to have skill too. Women are said to be more patient than men and, therefore, make good snipers. But we discovered that being a sniper is a difficult art and we mastered it only after months of daily lectures and training.'

One of the most famous of the Russian women snipers was Mila Pavlichenko from the Ukraine. She became so feared that the Germans referred to her as 'The Russian Valkyrie', after the mythical Norse maidens who visit the battlefield to escort the fallen to Valhalla.

Pavlichenko was a notoriously tenacious sniper who once survived a three-day sniper duel with a German near Sevastopol. She was hiding in a tree when a bullet hit just above her. Realising she had been spotted she fell to the ground as if she had been shot. Her adversary was an experienced sniper and not about to rush out and give away his position. Mila lay motionless for most of a day before the German was finally convinced she was dead. As he eventually came towards her she was able to grab her rifle, get him in her sights, and shoot him. After another hour of lying motionless she went to examine the body. Among his papers she found a small notebook. It was the 'kill book' kept by every sniper, cataloguing his activity. This German had been active since Dunkirk and had killed more than 400 British and French soldiers before Mila Pavlichenko added him to her total in the Crimea.

Mila Pavlichenko had joined a student sniper school while she was studying history at the University of Kiev. When the Germans invaded she volunteered and asked to be sent where her skills could be put to best use. 'What began as my hobby became my profession in the army,' she said. She eventually became so famous that she was invited to the Kremlin, then sent overseas to help encourage the other Allies to take the fight to Germany. Eleanor Roosevelt had invited her to America, and one speech in Chicago sums up her attitude.

'I am twenty-two years old,' she told a stunned audience, 'and I have already destroyed 309 enemy soldiers who have invaded my country. I hope you will not hide behind my back for too long.'

Although her comments seemed shocking at the time it is hard to believe that they were entirely spontaneous. It would have been a huge diplomatic affront otherwise. Pavlichenko seems more likely to have been an acceptable face of the European warrior spirit to present to the American people to reinforce Roosevelt's desire to support his European allies more forcefully.

The cult of 'sniperism', as the Soviets referred to it, meant that news of the snipers' exploits spread rapidly along the Front. They were lauded in Party propaganda newssheets, and Russian children could tell you how many 'Fritzes' each sniper had killed just as American children can quote baseball statistics.

Among the pantheon were the Okhlopkov brothers, Fyodor and Vasily, who were Yakuts from Siberia, by tradition the deadliest marksmen of the Soviet Union. Both were hunters who travelled thousands of miles each to enlist in the Red Army. Very early in the fighting Vasily was killed by a German sniper and, so the story goes, died in his brother's arms. Fyodor Okhlopkov promised his dead brother that he would avenge his death. Before he was wounded in 1944 and had to be withdrawn from the front line he had taken revenge 456 times in just over two years.

The archetypal Soviet sniper was Vasily Zaitsev, a shepherd from the Urals who had learned to shoot as a young boy when he hunted the wolves that threatened the family's livestock. Zaitsev was not the most successful of the Soviet snipers but he certainly became the best-known, thanks to Alexander

Scherbakov, head of the Red Army's political department. Scherbakov received daily reports of the fighting in Stalingrad in 1941 and appears to have decided that Zaitsev, the farmer turned killer, was just the hero that the Soviet Union was looking for. Zaitsev was known as 'The Hare', a crude translation of his name in Russian, and he attracted a crowd of young disciples who quickly became known as 'zaichata' or 'leverets'.

In the midst of the celebrations of the anniversary of the October Revolution it was announced that Zaitsev would take his total of kills to 150 in honour of this glorious event. Zaitsev seems to have been a modest and relatively unassuming man. It is likely that his projected tally was an invention of Scherbakov's propaganda machine rather than a genuine claim by the sniper. In any event by the end of October it was reported he had killed 149, one short of his tally.

Many snipers, especially lionised Soviets, are defined by the duels they fought and won. However, Vasily Zaitsev's name lives on because of a duel that probably never took place. Zaitsev was essentially a propaganda weapon. There is no doubt that he existed and that he was a skilled and deadly sniper, but the personal interest of the Red Army's political division means that Zaitsev's most celebrated encounter must be viewed sceptically.

The story goes that Zaitsev was proving so deadly to the Germans that they decided special measures were required. The Germans issued an order to 'find and kill the Russian hare'. The man who would execute that order was Major Konings, the senior instructor at the German sniper school at Zossen and a former big-game hunter. Konings came to Stalingrad and proved to be a deadly adversary. He killed so many Soviet snipers so quickly that it became obvious that he was some kind of 'super sniper'. It became even more obvious that, based on information from an interrogated prisoner, Konings was in Stalingrad with the express purpose of eliminating Zaitsev.

Konings carried the fight to Zaitsev by shooting two of his closest comrades. One, Morozov, was killed while the other, Sheykin, was seriously wounded. Zaitsev, thus provoked, decided that he was honour-bound to fight this duel on his own terms. He and his spotter, Kulikov, went out into the shattered ruins of the city to look for the German. Eventually they found him, lying in wait for another sniper. Zaitsev was preparing to shoot when a political officer who had gone along with him gave away their position, allowing Konings to shoot and to escape. His shot missed but the hunt was joined in earnest. Each knew the other was determined to make the killing shot.

They tracked each other until Zaitsev suspected that Konings had constructed a hide under a sheet of metal in no man's land, waiting for his shot. Borrowing a tactic from Hesketh-Pritchard, Zaitsev raised a mitten on the end of a piece of wood. The German was deceived by it and fired, allowing Zaitsev to work out the angle of the shot, and from that Koning's position.

The Soviet sniper's account of what happened after that has become sniper folklore.

After lunch our rifles were in the shade and the sun was shining directly onto the German's position. At the edge of the sheet of metal something was glittering: an odd bit of glass or telescopic sights? Kulikov carefully – as only the most experienced can do – began to raise his helmet. The German fired. For a fraction of a second Kulikov rose and screamed. The German believed he had finally got the Soviet sniper he had been hunting for four days, and half raised his head from beneath the sheet of metal. That was what I had been banking on. I took careful aim. The German's head fell back, and the telescopic sights of his rifle lay motionless, glistening in the sun.

It is a gloriously romantic story, just the sort of thing to boost the morale of the hard-pressed defenders of Stalingrad and urge them to even greater efforts. But while it sounds thrilling it doesn't sound true. For one thing no one is certain what the German sniper was called. In some versions of the story he is called Konings, in others he is called Thorvald, and in some he is not named at all other than as 'the Berlin super-sniper'. Moreover, in the incident leading to his death he appears to be astonishingly naïve for the best sniper in the Wehrmacht and the senior instructor at Zossen. Bad enough that he fell for the mitten-on-the-pole trick, but to give away his position without waiting to make sure his target was dead is a classic and usually fatal mistake.

The story stands or falls on two issues. Those who believe it's true will point to the telescopic sight which is on display in the Armed Forces Museum in Moscow. This, it is said, is the very sight that Zaitsev took from Konings' rifle. It is, however, indistinguishable from any other sight mounted on any other German Gewehr rifle. It could have come from anywhere. Those who disbelieve the story have their argument summed up by the historian Antony Beevor, who points out that nowhere is the incident mentioned in any of the daily reports to Scherbakov. Given that it was Scherbakov's job to promote Zaitsev and the cult of sniperism it is unthinkable that he would not have had it drawn to his attention.

Zaitsev's story was turned into a film called *Enemy at the Gates*. The British actor Jude Law played Zaitsev while the American Ed Harris was Konings. Despite extensive research, the director Jean-Jacques Annaud was unable to come up with any compelling evidence to support the facts. Instead he chose to think of the story as an allegory of these two great superpowers colliding in microcosm on the shattered streets of Stalingrad.

What is certain is that Zaitsev's epic fuelled Stalin's need for a hero. The warrior myth was vital to the survival of Stalingrad and with it the entire Soviet Union. Ironically, as its greatest example, the Zaitsev story calls into question the validity of all of the other totals attributed to Soviet snipers. Among those who doubted that many of them had reached three figures was Clifford Shore, who says:

After the war I questioned many Germans who fought on the Russian Front and asked often about sniping. I was told that there was actually little sniping on the Eastern Front. I have met some Russians who had been Red Army men

239

and saw them shoot in the summer of 1945. If their shooting prowess can be taken as a criterion, I think that the printed Russian figures of sniping casualties should be divided by a hundred.

Shore's version is as damning as Scherbakov's would be laudatory. The truth, as usual, will lie somewhere in the middle. Doubtless the totals of the Russian snipers were exaggerated by their political masters for propaganda purposes, but it seems unlikely this was done to the extent that Shore suggests. To say there was 'little sniping' on the Eastern Front seems equally groundless.

The totals may not have been as high as reported, but the Wehrmacht could and did attest to the impact of the Soviet snipers on the Eastern Front. The cult that grew up around them, along with the attendant hero worship, turned these lone-wolf killers into mythic, almost mystical, figures. Stories of their exploits would pass from street to street in a firestorm of whisper and rumour. The Soviets may have had no food and little ammunition, but as long as they had their snipers they had hope. And even if they had nothing more than that, they could still defend the Motherland.

Before Soviet propagandists turned their snipers into folk heroes the Russian Army had been on the bloody end of sniper duels in Finland in the Winter War. Almost sixty years later history seemed to be repeating itself. During the Chechen conflict stories began to emerge about a deadly sniper unit known as the White

Stockings. Tales of this elite force are as shrouded in uncertainty as the truth of the Zaitsev duel at Stalingrad. But like all potential myths it seized the imagination of the Russian troops based in Chechnya.

The White Stockings were all female, and all were champion biathletes from Ukraine. The biathlon is a winter sports event combining long-distance skiing with precision target shooting. Recognising that the women were among the deadliest snipers, the Chechen rebels were prepared to offer them vast sums of money to fight for them. They earned in a day what others would earn in a month, about 2,000 roubles. As far as the Russians were concerned, when they were pinned down by snipers in the streets of Grozny it was the White Stockings who were responsible for the biggest number of killings. Terrified soldiers would swear that they could hear these modern-day Amazons calling to them and taunting them, threatening to shoot off their manhood if they so much as showed themselves.

No one has ever admitted to being a White Stocking, or spoken to one, or even seen one. They are as elusive, it seems, as the Yeti. When they are captured they are allegedly, and conveniently, hanged in secret by the Russian authorities who do not want to make public martyrs of them. Only on one occasion did the Russians parade a scared, mousy-looking twenty-year-old known only as Anya, who they claimed was a White Stocking. She did not look like the sort of woman to instil fear into anyone, nor was there any evidence provided to support their claim.

In October 2001, the *Moscow Times* found another woman who it claimed was determined to go to

Chechnya and be a Russian equivalent of a White Stocking. Galina Synitsina, a former champion shooter, had lost her job, and to support her family wanted to enlist with the Russian Army as a contract sniper. The Russians used contract snipers in Chechnya but refused to take Synitsina because, at forty, they claimed she was too old. Other contract snipers who signed on quickly signed off again when they found out the wages were not as good as they had been promised.

The White Stockings remain a mystery, a battlefield folk tale. And when the fighting ceased? They stole away of course, back to where they had come from, never to be heard of again.

14

FOUR KILLS FOR A DOLLAR

After VE Day, once again British and American high commands showed no instinct to heed the lessons of history. Just as they had in 1918, both decommissioned their specialist sniper regiments. Barely five years passed before they needed them once more.

The conflict between North and South Korea reflected in many ways the battlefield conditions of the Western Front. Two armies lined up across a border stretching across the Korean Peninsula for a little over 150 miles. Two essentially static armies faced each other in a war of thrust and counter-thrust that on the face of it seems tailor-made for snipers. Certainly the Chinese who were fighting on the side of the Communist North made some use of snipers, but South Korea's American and British Allies had none.

The more enlightened American commanders quickly designated the best shots in their units as snipers and set them to work. They used whatever contacts they had to get these men the best guns and telescopic sights available.

A more formal counter-sniping tactic evolved in 1952, two years into the war. An account of the incident that was said to have provoked this decision first appeared in the *Marine Corps Gazette* in 1963, some ten years after the Korean War had ended. The *Gazette* reported that on his first day on the line the new, unnamed, commanding officer of the 3rd Battalion, 1st Marines, stepped out on to a hillside to review the situation and had his binoculars shot out of his hands by a North Korean sniper.

> The battalion commander was only scratched but he reflected that this was a helluva situation when the CO could not even take a look at the ground he was defending without getting shot at. Right then and there he decided that something had to be done about that enemy sniper ... As he made his way back from the outpost, his mind mulled over the problem of enemy snipers. A decision was made before he reached his command post. He would form a sniper unit.

> When he found out that he had a good supply of rifles with telescopic sights as well as infrared sights for night fighting, the colonel then sought out a veteran gunnery sergeant who had trained rifle teams. He told the sergeant what he had in mind, and the sergeant selected volunteers from the best shots in the battalion and put them through an intensive three-week course. By the time they were finished they were able to take up positions in camouflaged bunkers from which they became highly effective counter-snipers.

> 'In nothing flat there was no more sniping on our

position,' the colonel later recalled. 'Nothing moved out there but what we hit it.'

As Hesketh-Pritchard had done in 1915, these snipers then spread out to other units to pass on their expertise. British and Commonwealth troops serving in Korea also formed their own ad hoc training schemes to put sniper units in the field. Yet although these measures were effective, none of them had the official blessing of their respective military establishments. The Korean War ended with the signing of a treaty in July 1953, and when the shooting stopped there was still no formal sniper training programme in any of the Allied armies.

After the Korean War, sniping was again relegated to the darker recesses of military thinking. As on the Western Front, sniping in Korea was recognised only as a local solution to a specific problem. Although training in marksmanship continued, the skills of field craft, intelligence-gathering and camouflage, which the sniper needs if he is to be genuinely effective, were neglected. Most divisions maintained shooting teams which competed for marksmanship trophies, but by the late 1950s there was even a move to disband these as a waste of time and resources.

Salvation came in the form of the US Marine Corps shooting team based in Hawaii. One of its members, Chief Warrant Officer Charles Terry, was far-sighted enough to realise that the team faced being broken up and its skills and men lost to the service. Terry voiced his concerns to Lieutenant Jim Land, who was the officer in charge of the shooting team. Terry warned Land that if they didn't provide value for money as a rifle and pistol team then they would be broken up.

'They're not going to pay for us to run around the country and shoot – we have to deliver something worth the money . . . we might even give the team a new meaning by pushing the sniper angle.'

Land recognised the merit of what Terry was saying. He did some research into the history of sniping, checked that the equipment was still available, and presented a paper to his superior officers. They were impressed, and towards the end of 1960 Land began a two-week sniper course at the Puuloa Rifle Range. Land's course – the first week was marksmanship, the second field craft – remained the only formal sniper training in the American Army for several years. It was the sort of officially unofficial set-up that Hesketh-Pritchard would have recognised from his own battle for acceptance in 1915. There are few records of its existence, a fact that is confirmed in a press release issued by the Marines in January 1962 saying just that. However, perhaps trying to put some sort of spin on what it obviously perceived as a non-story, the 1st Marine Brigade Informational Services Office did include one significant paragraph:

In this age of push-button warfare, little thought is given to the common infantryman who has nothing but a ten pound rifle and a lot of courage. But beware of the sniper – he is deadly.

Deadly he may have been, but the military establishment seemed to have concluded that the sniper's time had passed. In an age of intercontinental ballistic missiles the world stood on the brink of mutually assured nuclear destruction. What could even the best sniper offer against an opponent who launches an attack from a silo thousands of miles away? What sort of deterrent was one shot, one kill when one missile could obliterate thousands? But at the moment when sniping seemed sure to be rendered obsolete, America undertook another war against an almost pre-industrial society. For all the technological prowess of the USA, the sniper would be one of its best and most cost-effective weapons.

The Marines of the 3rd Battalion, 9th Regiment stepped ashore in Vietnam on 8 March 1965. They were the first regular American troops to enter South Vietnam, there ostensibly to protect the American air base at Da Nang. Over the next five years they would be joined by more than half a million of their colleagues: by 1968 there were 536,000 American troops in Vietnam.

As these Americans began to pour into South East Asia it became apparent that not one of them had been trained as a sniper. All the ordnance, heavy artillery and air strikes that could be called in at a moment's notice were of limited effectiveness in what was essentially a jungle war. The conditions were perfect for snipers; the Americans had none. Some soldiers tried to make the best of the situation by asking their families to send them telescopic sights or buying the scopes themselves. These served a purpose, but they were far from perfect for jungle conditions. The US Army attitude to sniping can be gauged from the fact that the Field Manual for care of rifles that was issued in May 1965, only weeks after the Marines had arrived in Vietnam, made no mention of sniper rifles. Eventually pressure from commanders in the field who kept requesting sniping equipment forced the Army to establish a sniper-training programme in August 1965. The programme was set up under Major Robert Russell, a veteran of three wars and a man with a 22-year-long career on Army shooting teams. Marine Staff Sergeant Don Barker is generally credited with the first American sniper kill in Vietnam.

One of Robert Russell's first instructors was Jim Land, by now a captain, who had set up the training programme on Hawaii in 1960. It was his job, amongst other things, to welcome the volunteers who had been selected for his training programme. The selection regime was rigorous and marksmanship was only part of the process. Land's set-piece speech was rousing, without sugar coating what the volunteers would face. He told them they had not been chosen because they were meaner or tougher or better than anyone else. They were selected first and foremost because they were good Marines with a good sense of values and strong moral fibre including, ironically, the desire to hold life sacred. These were attributes, according to Land, which were vital to be a successful sniper.

When you go on a mission there is no crowd to applaud you – no one for whom you can flex your muscles or show how tough you are. When you go on a mission you are alone.

You have to be strong enough to endure physically lying in the weeds day after day, letting the bugs crawl over you and bite you, letting the sun cook you and the rain boil you. Shitting and pissing in your pants but lying there. Lying there because you know Charlie's coming and you're gonna kill him.

You don't select the first gooner that comes into your field of fire either. You select your target carefully, making sure that the gooner you kill is Charlie, so you can waste the bastard with no doubts or remorse . . . I know that as grunts it was easy for you to feel justified in killing the enemy when he attacked you – he was trying to kill you. If you attacked him he also had a choice to fight or surrender – you did not murder him because he died trying to kill you. That's self-defence.

As a sniper you do not have that luxury. You will be killing the enemy when he is unaware of your presence. You will be assassinating him without giving him the option to run or fight, surrender or die. You will be, in a sense, committing murder on him – premeditated.

To deal with this successfully you must be mentally strong. You must believe in what you are doing – that these efforts are defeating our enemy and that your selected kills of their leaders and key personnel are preventing death and carnage that this enemy would otherwise bring upon your brothers.

It was a powerful speech that left his audience under no illusion of what would be expected of them.

It was also very seductive. Land's audiences were young men, many of them barely into their twenties. In the early stages of the war, most were volunteers rather than draftees. Theirs was a generation brought up on the moral values of John Wayne films. Many of them saw themselves in Vietnam as reliving Wayne's screen exploits in films such as *Sands of Iwo Jima*. Land was giving them the chance to live their dream in all its awful and dubious splendour

The reaction to the conclusion of his speech was almost always the same. There would be a few seconds of contemplative silence, then his audience would take the roof off, whistling and cheering and yelling.

One of those who had heard Land's speech many times was Gunnery Sergeant Carlos Hathcock, who had been hand-picked by Land to join his team. Hathcock was a national shooting champion who had been serving as a military policeman in Chu Lai, a major American base some 50 miles south of Da Nang, when Land arrived in the country. He became one of the small group of men who would form the basis of Land's sniper-training programme.

Hathcock was an unassuming man. Married with a child, his life seemed to be dedicated to the Corps. In his tour of duty in Vietnam he became a legend. Hathcock recorded 93 confirmed kills, although he probably accounted for many more. A kill was not official until a sniper had confirmed it by examining the body and then entering the details in his 'kill book'. Although he always insisted that the statistics were meaningless, Hathcock estimated he killed half as many again as he is officially credited with.

His shooting skills were extraordinary, whether with a rifle or a larger gun. He once shot a Viet Cong at 2,500 yards with a single bullet from a .50 calibre machine gun. A consummate professional, he attracted the highest praise from Jim Land:

The sniper does not hate the enemy; he respects him or her as a quarry. Psychologically, the only motive that will sustain the sniper is the knowledge that he is doing a necessary job and the confidence that he is the best person to do it. On the battlefield hate will destroy any man – and the sniper quicker than most.

The sniper is the big-game hunter of the battlefield and he needs all of the skills of a woodsman, marksman, hunter and poacher. He must possess the field craft to be able to position himself for a killing shot, and he must be able to effectively place a single bullet into his intended target.

Gunnery Sergeant Hathcock was all of these things with a full measure of the silent courage and quiet optimism of a true champion.

Like the greatest of his profession Hathcock lived by the sniper's creed of one shot, one kill. He was a natural hunter who would spend days on the track of his quarry. He could kill men or women as the occasion demanded. In Vietnam, for the first time, snipers found themselves shooting children used by the Viet Cong to plant booby traps. The age or gender of his target was immaterial to Hathcock, who fulfilled all of his missions with clinical detachment. He was also called upon to carry out what amounted to political assassinations. On one occasion he was asked to kill a French civilian. Hathcock suggested that the Vietnamese should arrest him but he was told that this man was helping the Viet Cong and had to be killed and killed quickly. Hathcock was flown by helicopter so far into the jungle that he barely knew where he was, and when his target approached down a jungle path he took him out with a single shot. He was later told that the Frenchman had been helping interrogate prisoners for the North Vietnamese. Whatever the Frenchman's crimes – and they may have included the torture of captives – Hathcock's mission was an act of sniping as terrorism. But he did not flinch. He analysed his kills obsessively, and always avoided routine since he knew that committing himself to a set pattern was the surest way of being killed by an enemy sniper. It was a competition, he constantly reminded himself, and second prize was a body bag.

Hathcock was a natural, a man to whom the discipline and challenge of sniping came easily, and a man who prized his ability to kill with a single shot. His victims never knew what had hit them.

The sniper will always go for the head shot, the instant kill. The trained sniper, like Hathcock, will also stack the odds in his favour by aiming for what is known as the 'triangle of opportunity'. The sniper wants to place himself on high ground or in a sufficiently elevated position to be shooting down on his target. If he misses the clean head shot then the bullet should still travel through the head, down through the neck, along the long axis of the body, through the heart and out of the back. This vastly increases the chances of hitting a major organ such as the heart or liver or severing an artery. In either case death is almost inevitable.

Civil War surgeons described the effect of a Minié ball as it shattered long bones, driving them into tissue and turning them into mincemeat. The situation with a high-velocity bullet is even more dramatic.

Hugh Thomas is a retired surgeon who, for five years, treated gunshot wounds and sniper wounds in Northern Ireland. He is now an expert forensic witness in the field of wound ballistics.

Think of the bullet as being like Concorde in flight, in that it drags a vacuum behind it. A good analogy would be a shuttlecock where the bullet is the head of the shuttlecock and the vacuum is the feathers; those dimensions are roughly the same in terms of relative size. The initial entry hole creates a ballooning cone of kinetic energy. This cone looks like the photographs you will have seen of a plane breaking the sound barrier where you have a bow wave that extends up and over the nose of the plane. This cone will create a cavity which is about two and a half times the diameter of the bullet, and it will taper down quickly to about three or four times the length of the bullet.

If it strikes a solid organ like the liver or a kidney or the spleen then there is huge destruction of tissue as the bullet passes through coupled with the cavitation effect of the immediate impact. This can cause a substantial hole perhaps up to two inches in diameter. The cavitation in fact creates a vacuum that sucks material into the wound, including cloth and metal which can cause infection. Nature hates a vacuum so the body rushes to fill it with whatever is there; blood, bile, body fluids, which can create sepsis if the patient isn't treated in time. It all happens in a millisecond but the effects are very dramatic. As a surgeon I once treated a man who had been shot with a high-velocity bullet that travelled downward through the liver, the lower torso, the leg and eventually came out at the knee. I was finding liver puree coming out of the exit wound behind this man's knee.

High-velocity bullets are particularly bad for the brain, the cavitation effect purees whatever is in there making recovery almost impossible.

Whether Carlos Hathcock knew or even cared to know about cavitation or cones of kinetic energy is a point of conjecture. A private man, Hathcock kept his feelings to himself. He does not appear to have had any great respect for the Viet Cong, referring to them as either 'hamburgers' or 'hot dogs'. Equally he does not appear to have been motivated by any great personal crusade or sense of outrage. Sniping was a job, and it was one at which he happened to excel, as he once explained to a superior officer:

I like shooting, and I love hunting. But I never did enjoy killing anybody. It's my job. If I don't get those bastards, then they are gonna kill a lot of those kids dressed up like Marines. That's the way I look at it . . . But I never went on any mission with anything in mind other than winning this war and keeping those . . . bastards from killing more Americans.

Hathcock once showed emotion when he stalked and killed a Viet Cong woman who had been responsible for torturing young GIs to death. Her tactic was to kill them slowly and painfully so their screams could be heard by their comrades. Hathcock and Jim Land went into the jungle and hunted the woman down. Hathcock killed her with a single long-range shot as she was trying to make her escape. He then fired another high-powered bullet into her body as she lay on the dirt. When he was sure she was dead Hathcock smiled and pounded his fist on the ground in triumph. It was the only time he allowed himself to register any emotion over a kill.

Carlos Hathcock was not the most successful American sniper in Vietnam; that distinction is held by Army Sergeant Chuck Mawhinney, an unprepossessing man who is credited with 101 confirmed kills. However Hathcock did have a knack for attracting attention. He wore a white feather in the band of his bush hat which made him stand out from the other snipers. The Viet Cong knew this only too well. There was a bounty on any US sniper killed or captured. For run-of-the-mill soldiers the bounty was about the equivalent of eight American dollars; for Long Tr'ang du K'ich – 'White Feather Sniper' – the reward ran to more than ten thousand dollars. Hathcock and Jim Land both featured on wanted posters which were distributed extensively throughout the Vietnamese villages.

Although he was notorious among the enemy, Carlos Hathcock led a charmed life. He came close to death in the jungle on many occasions but survived. One of his closest calls came when he and his spotter pursued a Viet Cong sniper through the dense jungle in a classic counter-sniper hunt. The Vietnamese sniper was a skilful and versatile opponent and it took Hathcock a long time to track him down. Just when he thought he had lost his quarry Hathcock spotted the smallest glint of reflected sunlight. He decided to gamble on a shot, took careful aim, squeezed the trigger, and fired. When he and his partner found the body it emerged that Hathcock had shot the Vietnamese sniper through his telescopic sight, the bullet had shattered the scope and flown straight into his brain through his eye. After being congratulated on the skill of the shot, Hathcock was struck by a chilling thought. The only way he could shoot a man like this was if the man had been pointing the gun directly at him. The glint of light he had seen was the sun reflecting on the Viet Cong's sniper scope; Hathcock had fired fractions of a second before the Vietnamese sniper would have shot him.

Stories like this made Carlos Hathcock a man to be admired and respected by his fellow US soldiers, although some found the coolness with which he killed inhumane. However Hathcock is best remembered for a single act of heroism that owed nothing to his skills as a marksman. He was riding in a half-track armoured vehicle with some other Marines when it was shelled in an ambush and caught fire. Although he was on fire himself, Hathcock stayed inside the burning wreckage and threw his injured comrades to safety. Every soldier who made it out of the vehicle owed his life to the Gunnery Sergeant.

Hathcock eventually staggered out of the truck himself, burning from head to foot. He was burned over most of the surface of his body with full-thickness burns over 40 per cent of his body. Had it not been for the fact that his lungs were still unscarred and intact he would almost certainly have died. As it was he spent more than a year recovering from his injuries. During his stay in hospital he was also diagnosed with a form of multiple sclerosis. Hathcock's active career as a sniper was over but he continued to pass on his knowledge as an instructor. He was one of the first appointed to the Marine Corps Scout/Sniper Instructor School that began operating at Quantico in 1977. It was the American military's first permanent sniper training establishment.

Hathcock's reputation preceded him wherever he went in Vietnam. On one occasion when he and a fellow sniper came into a mess tent, another soldier shouted 'Here comes Murder Incorporated.' Hathcock tried not to react to this kind of provocation, but there were those who revelled in their unsavoury tag. There was, for example, a sign at the Marine sniper school at Da Nang which announced to the world

War Our Business, Death Our Only Product

When the Scout/Sniper Training School was established at Quantico it also boasted a defiant legend on a sign outside:

The average rounds expended per kill in Vietnam with the M-16 was 50,000. Snipers averaged 1.3 rounds. The cost difference was $2300 vs. 27 cents.

So with roughly four Viet Cong kills for a dollar the sniper not only made good tactical sense, he made good economic sense too. He was a cost-effective killer, an economic reality whose consequences would soon be evident on the world stage beyond the formal theatres of war.

Snipers have always been drawn to their job partly by their independence from the command structure. They were generally excused regular duties, but nowhere did they enjoy the liberty they had in Vietnam. Snipers could come and go almost as they pleased and no one would question them. It's not surprising that the life of the sniper, despite Jim Land's unvarnished welcome speech, was still an enticing one for young recruits, especially those who had signed up in the early days of the war. They saw sniping as a means of fighting for their country and making a difference on a one-to-one basis.

Joseph Ward was a young man from Longmont, Colorado who joined the Marines after he left high school. He was eighteen when he signed up in August 1968. The following April he found himself in Vietnam as a scout sniper partnered with the famous Chuck Mawhinney. He had been in Vietnam for barely a week when he went out on his first sniping mission. His platoon had come across a unit of NVA regulars in the jungle. When the call 'Snipers up' came back from the point man Mawhinney started to race forward with Ward close behind.

The lieutenant was yelling and swearing as loud as he dared for Chuck to slow down and watch out for booby traps. If Chuck heard him, he didn't give any indication, and in a few seconds he assumed a kneeling position. As trained I took my spot two feet to his left rear. By the time I focused my field glasses and confirmed that they had weapons, Chuck had his breathing under control and the first shot rang out.

A VC at the centre of the column dropped and I heard myself say 'Hit!' the same way I had said so many times at the rifle range. It came out of my mouth calmly, well rehearsed. All the VC, all but one, began dragging their fallen comrade towards the tree line they had come from. A second shot broke the early morning quiet and a man toward the front of the retreating column fell. 'Hit,' I said as I watched them pick up the second body and make for the trees . . . Another shot rang out, and he [a third Viet Cong] fell behind a rice paddy dyke. For the third time in barely half a minute, the word 'hit' came from my mouth. I felt nothing. The reality of the situation hadn't sunk in yet.

Some days later, when Ward and Mawhinney returned from their trip into the bush, Ward wrote to his mother about the incident:

I have three confirmed kills already. It's not like the movies when a GI kills his first enemy soldier. It wasn't as bad as I thought it would be, especially after I've seen what they can do to us. If people back in the States could see how dangerous Communism is and how fortunate they are to have what they do, there would be fewer riots and protests. The guys over here don't pay much attention to the protesters. I've also developed a real dislike for racists. Black men bleed just like white men. The colour of a person's skin means nothing.

Ward's trip with Mawhinney was the older man's last sortie as a patrol leader. He served out the remainder of his time at headquarters. Before he left he entrusted to Ward the rifle that had seen him through his tour of duty. The young sniper, who went on to use the gun with deadly effect himself, says it was in such pristine condition it could have gone on sale in a sporting goods store without anyone ever knowing it had been used.

One of the most important jobs for the scout sniper was to confirm the kill. This involved finding and identifying the body where possible as well as searching it for maps, papers, or other documents. The good sniper could tell a great deal by the condition of his victim. Was he or she well fed? Were they carrying ammunition? Were their uniforms in good condition? This information would be relayed to Military Intelligence to be interpreted.

Ed Kugler served for two years as a Marine sniper in Vietnam. His memory of his first kill is typical. He described squeezing the trigger of his rifle on the firing range as a 'natural high', indeed, his instructor had told him to compare it to fondling his girlfriend's breast. However, once he qualified the highs were in short supply, as he and his partner, Zulu, spent weeks of drudgery in the jungle without ever firing a shot. Finally his chance arrived:

We got our break early in week three. I spotted some bad guys to the west . . . in a place we hadn't seen activity before. To that point, Zulu and I had been working on the war by calling in artillery. We were getting a few gooks with the big guns but had done no real sniping. The lieutenant came and took a look. He asked whether we could shoot that far. Why was this guy questioning us? All he had to do was give us the word and these little assholes were history . . . It looked to be like a seven-hundred-yard shot. We had to guess how the angle of shooting from so high above the target would affect our aim. We checked the wind by watching the tree tops between the target and us. The wind seemed to be in our favour. It was still morning, about 1000 hours, so heat wasn't yet a factor . . . The ship's scope confirmed the uniformed and ugly guys at seven hundred metres. One had a rifle and that's about all we could tell. I wanted their asses and I wanted them bad. Zulu spotted for me. It was a tough shot from so high up. I couldn't lie flat, so it was a shot from a modified sitting position. I sat, twisted nearly straight up, with my rifle resting on my pack. I took my time until I had him squarely in the crosshairs. This was it, man . . . relax . . . breathe in, you can do it, be gentle . . . breathe out halfway . . . hold it . . . hold it. Rock him in your sight . . . okay. The trigger . . . you're squeezin' your girlfriend's tit . . . gentle now . . . here it comes . . . BOOM! I squeezed off a round from my 30–06. 'Got his ass, Kug!' Zulu yelled. 'Helluva shot.'

By the end of the day Kugler had shot three Viet Cong and his partner, Zulu, two. In his memoir Kugler described it as 'a good day for snipers after a long drought'.

In more recent interviews Kugler would describe that first kill as being like a touchdown in terms of the exhilaration he felt. But other than that, like Hathcock and many of his comrades, Kugler says he did not agonise long if at all over how he earned his combat pay.

'I didn't have any feeling. In fact I worked at not having any feelings and I paid for that for years afterwards. There was no remorse; there wasn't anything.'

As Kugler hints, the price for being a sniper can be a high one paid either in years of recurring nightmares, or in the sort of emotional repression that will blight normal relationships. Perhaps this is what Jim Land meant in his belief that being a sniper required what he called 'a special kind of courage'. Not the fight-or-flight response that will make a soldier charge an enemy position – it was the exact opposite of that adrenalin-fuelled surge of bravery. Impossible to teach or to learn through training, it was innate. It was the nerve and willpower to allow the sniper to be alone with his thoughts, his fears, and his doubts. And still pull the trigger.

Ed Kugler fitted Jim Land's description of a sniper. He would not allow emotion to intrude on his job any more than a duck hunter would concern himself with the fate of the duck. Kugler, like Mawhinney and Hathcock and all the other US snipers, believed he was doing a necessary job – Joseph Ward believed he was personally stemming the flood of Communism. If they could not continue to convince themselves of that then they could not shoot. But what strength of will must have been required to maintain that conviction and that lack of emotion.

The sniper must be able to calmly and deliberately kill targets that may not pose an immediate threat to him. It is much easier to kill in self-defence or in the defence of others than it is to kill without apparent provocation. The sniper must not be susceptible to emotions such as anxiety or remorse. Candidates whose motivation towards sniper training rests mainly in the desire for prestige may not be capable of the cold rationality the sniper's job requires.

That description from Hesketh-Pritchard is the key to the craft of the sniper. It is a craft without feelings, based on rationality and hard logic. The essence of the sniper's training is to turn him into a professional killer in the strictest sense of the word; he is a man who exists simply to shoot other men and women – and occasionally children. His actions must be governed by an overriding imperative, the sense that what he is doing serves a higher authority than his own. In many ways snipers are like hangmen or public executioners: it is a job that, if it has to be done, must be done at the absolute highest level of ability.

The psychological toll on the sniper as he carries out his art – and sniping is more art than science – is massive. Snipers work in pairs for sound tactical reasons, but there is an equally sound psychological reason for the sniper team: the presence of another man can ease the enormous mental pressure of the job. No matter how well trained a sniper might be, killing is hard; it goes against nature, and to have someone there to either share the blame or ease the moral burden can be invaluable.

It had always been assumed that men would kill without compunction in the heat of battle, if not because they were fighting for their country or their family or their honour, then for the simple fact that someone was firing at them. This belief was questioned by the military historian Brigadier General S.L.A. Marshall during the Second World War. He asked around 400 men who were fighting in the Pacific Theatre what they had done during a battle. He was astonished to discover that out of every hundred men who took part in the action, no more than fifteen or twenty had fired their weapon. Marshall's criteria for firing were fairly generous. It was not necessary to aim or indeed to hit anything; all you had to do was point the gun and shoot in the direction of the enemy. Even so, whether the action was over in minutes or lasted for days, no more than one man in five fired.

Marshall's methodology was later questioned, but

his broad conclusions were confirmed by the anecdotal evidence from officers in the field. Combat NCOs would admit that it was almost impossible to get a man to fire his weapon unless you physically stood over him and badgered him into shooting. In the First World War a doctrine of live and let live had been actively practised in various parts of the line, which amounted to opposing troops mutually agreeing not to shoot at each other, or if they did then they would fire harmlessly into the air. It was not uncommon for officers to have to go along the line beating soldiers with the flat of their sword to encourage them to aim straight. These were the same officers who, as the war wore on, had to order their men over the top at gunpoint.

The sniper cannot allow himself the luxury of not firing his weapon. Killing is his job, and either driven by the rightness of his cause, or simply by coldly repressing any emotion, he must be able to fulfil it. Snipers, especially those operating in Vietnam, would blackly boast that they felt 'nothing but the recoil'. Certainly men like Ed Kugler appear to have rejoiced in their mounting tally of kills and have told their stories in a boastful, self-aggrandising way. For some like Joseph Ward the first kill produced a feeling of numbness, because it was all too much to take in at the time. For those such as Marine Gary White the feelings were entirely different.

Through my scope I could see the five [Viet Cong] were carrying heavy packs and wore a mixture of uniforms and civilian dress. I braced my rifle and squeezed off a shot. By the time I regained the sight picture I could see my target going down. I swung around to try and get another shot but the other gooks quickly melted into the surrounding jungle. We moved down the hill and found the body and a pack full of medical supplies and rice. One shot, one kill. Man I was excited, completely exhilarated. It was like hitting a home run.

This initial gung-ho sensibility soon gave way to monotonous reality. Dave Nelson became a sniper because he felt it exemplified his notion of manhood and a warrior ethic. It was a pure, clean activity:

A clean hit was an accomplishment . . . I chose who lived or died because I looked through the scope and pulled the trigger.

Nelson had seventy-two confirmed kills in his two tours of duty in Vietnam, but on his second stint he began to lose his appetite for killing. Where once he had struck like a wrathful god, now he faced uncertainty and insecurity. He would, for example, tie himself to whatever tree he was hiding in to force himself to keep on fighting even if he was wounded. Uncharitably we might also suggest that this was to combat any, entirely understandable, failure of nerve.

'Once you've accepted your own death you can become really proficient at killing because it is no longer important if you die.' Nelson's Zen-like acceptance of his role in the scheme of things helped him to cope. For others it was the conditioning of their training that got them through. It was a job, and once the job was done there was no need to dwell on it. Chuck Mawhinney, the deadliest sniper in Vietnam, simply turned in his gun and walked away when his stint was over. He had done his duty.

'It was over,' he said in an interview twenty years later. 'I just wanted to forget about it and get on with my life. I don't spend much time thinking about Vietnam but the memories that I have are mostly positive.'

If that is the case then Mawhinney must count himself a lucky man. There are others with less positive memories of their time in the jungle, such as former sniper Thomas Ferran:

> Sniping is a very personal experience. You look through your scope and you see what the person looks like. You realise they are human. You've got complete control over their life, you've become God-like, in that once you pull that trigger and take them out – you just owned their life. As a sniper you carry that 'last sight picture' with you for the rest of your life.

It is perhaps this sort of knowledge which means the ordinary soldier still is not entirely comfortable in the presence of the sniper, a man whose only function is death. This is the sort of mistrust expressed by those of his colleagues who referred to Hesketh-Pritchard as 'The Professional Assassin'.

It was not the Viet Cong or an enemy sniper who put an end to Carlos Hathcock's remarkable tour of duty. His military career came to an end on 20 April 1979, some six years after the last American serviceman had left Vietnam. The injuries from his heroics in the fire plus the effects of his multiple sclerosis had combined to put him on the Corps' Disabled Retired List. The day before he left the service he was still teaching, still drumming into his students that the most deadly thing on the battlefield was a well-aimed shot.

As a retirement presentation he was given a commemorative rifle which had been made at the Marksman Training Unit's Armoury. The rifle bore a plaque with the following inscription:

There have been many Marines. There have been many Marksmen. But there has been only one Sniper. Gunnery Sergeant Carlos N. Hathcock. One Shot – One Kill.

15

NOTHING BUT THE RECOIL

The most famous sniper attack in history took place in Dallas, Texas around 12.30 on Friday, 22 November 1963. The victim was the President of the United States, John F. Kennedy. In the space of a few seconds his assassin, Lee Harvey Oswald, would introduce to a new generation the fear and panic their fathers had known in combat when faced with the unseen, unknowable sniper. Unlike the threat their fathers had faced – a highly disciplined, often highly motivated, soldier – the world would now have to deal with men who were rogue killers, often hired guns, who operated outside any army rules of engagement.

After the Kennedy shooting police found an Italian-made 6.5mm 91/38 model Mannlicher-Carcano bolt-action rifle with a telescopic sight on the sixth floor of the Texas Book Depository. It was a military surplus weapon originally used by the Italian Army. Oswald had purchased the gun by mail order from a Chicago sporting goods store. The vantage point was a classic sniper location, above and behind the target and well placed to take advantage of the triangle of

269

opportunity. Killing the President would not be difficult for a trained marksman. The shots were made at a distance of around 180 feet, well within the range of the rifle's iron sights, although a telescopic sight would have improved their accuracy.

The Warren Commission, set up to investigate the shooting, found that Lee Harvey Oswald had acted alone and that there had been three shots. This tallied with the three empty cartridge cases found in the Book Depository. One shot struck Kennedy, another struck Kennedy and went on to hit Texas governor John Connally, who was also in the President's car, and a third appears to have missed. The findings of the Commission did nothing to dampen the conspiracy theories that had sprung up in the wake of the assassination, indeed from the moment Lee Harvey Oswald had first declared immediately after his arrest: 'I'm just a patsy.' Some of these theories are more plausible than others; none is supported by any hard evidence. What evidence exists can be used to support the Warren findings as easily as to gainsay them.

Oswald was a Marine-trained marksman. He served with the Corps from 24 September 1956 until 11 July 1959, when he received a compassionate release. On 21 December 1956 he scored 212 in his marksmanship test, above average, and enough to be officially classed as 'sharpshooter'. Using a bolt-action rifle it would be difficult but not impossible for him to fire three shots in a little less than six seconds. While the Commission arrived at a consensus that there were shots, it conceded however that some witnesses heard only two while others heard as many as five or six. The discrepancy is attributed to panic, shock, distance, and the general acoustic landscape.

The most common conspiracy theory holds that there was a second gunman firing from a grassy knoll in Dealey Plaza. Oswald, so the theory goes, fired the first shot but the second, devastating, head shot came from the knoll. Those who are blamed for the assassination include the Mafia, the military-industrial complex, and Kennedy's political opponents.

One person knew for sure: Lee Harvey Oswald took his secrets to the grave. He was fatally shot two days later by Jack Ruby – a killing shockingly captured on national television – thus sparking a whole new round of murky conjecture.

Whether he acted alone or in concert, the immediacy of television meant that Oswald's genuinely were the shots heard around the world. A single gunman had set in train a series of events whose implications would be felt far beyond his immediate actions. All for the price of a couple of 27 cent bullets.

Would Kennedy have been as adventurous in Vietnam as his successor Lyndon Johnson? Would his survival have ushered in a Democratic dynasty with his brothers Robert and Edward following in his footsteps into the Oval Office? Would America's successive foreign policy initiatives have been any different?

In only one respect did Lee Harvey Oswald fail to live up to the credo of the sniper: he needed two shots. Nevertheless, he was the first to move sniping from the battlefield into daily life in America.

A no less clinical example of peacetime assassination by a sniper occurred, also in Texas, three years after

Kennedy's death. Just before noon on 1 August 1966 Charles Whitman went on to the University of Texas campus in Austin. He worked at the university as a lab assistant, so neither he nor the bulky gear he was hauling around with him aroused suspicion. A guard even issued him with a temporary loading permit to allow him to go across to the University of Texas Tower and park there while he gathered a few more things together. In overalls and pushing a two-wheeled dolly, Whitman looked as if he was beginning a regular day's work as he wheeled the laden dolly into the building.

Once inside he took the lift to the 27th floor and then dragged his gear up a short flight of stairs to the observation platform on the floor above. He arrived a little before noon. Edna Townsley, the observation deck receptionist, was due to finish her shift at midday. There was no one around as Whitman walked across to her desk and bludgeoned her on the back of the head with the butt of a rifle. As she fell he hit her again on the head. Edna Townsley survived the initial assault but died some hours later.

Whitman then passed a couple who, in retrospect, must have considered themselves the luckiest people in the world. Cheryl Botts and Don Walden walked into reception and noticed Whitman there with a shotgun in one hand and a rifle in the other. They also saw 'stuff' on the floor – this later turned out to be Edna Townsley's blood. Neither Botts nor Walden thought the situation sufficiently strange to comment on it, so they got in the elevator and went downstairs.

Another group of people, including two children, came up to the observation floor. They found the way barred by a desk. When two of them tried to shift it, Whitman ran towards them firing his sawn-off shotgun. One man died instantly, two more were critically injured, and others received injuries of varying degrees. Whitman's ninety-minute murder spree had begun.

The 25-year-old Whitman had prepared meticulously. A former Marine, he had his old service footlocker with him. Inside there was food, water, petrol, ropes, knives, ammunition and other supplies to see him through what he planned to be a long stint.

The observation deck of the 307-feet-tall tower gave Whitman a perfect field of fire. He was able to move around unhindered and could spray shots wherever he liked. For the next hour and a half, Whitman callously and cold-bloodedly sniped at anything that took his fancy. He simply raised himself above the parapet and fired. His victims included a pregnant teenager who survived herself, but the baby she was carrying died; a physics professor, a father of six, two high school students, and many others. Some of his shooting was of textbook efficiency: one young man was killed by a shot that went between two spars of a railing barely six inches apart. Whitman was 300 feet up and several hundred yards away.

Eventually the police arrived to find a scene resembling a battlefield. As the August heat approached 100 degrees, dead and wounded lay all around the base of the Tower. Whitman meanwhile still had the run of the building. Eventually three policemen led a group who started to make a move towards the observation deck. They moved up floor by floor, never knowing what waited them round the next stairwell. Finally they reached the 27th floor, where some of the group managed to evacuate the wounded from Whitman's initial rampage. Three policemen – Jerry Day, Houston McCoy and Ramiro Martinez – along with an armed civilian, Allen Crum, tried to make their way on to the observation deck. The door was wedged closed, but Martinez kicked and battered at it until the maintenance dolly that had been blocking it fell away. Martinez and McCoy crawled out on to the deck while Day and Crum guarded the door. As Whitman moved to a new position, Crum took a shot but the gun misfired. Nevertheless the noise of the shot moved Whitman back towards the other policemen. Ramiro Martinez rounded the corner towards Whitman, blazing away with his police issue .38. Whitman tried to shoot back but couldn't turn properly. McCoy meanwhile, coming from behind Martinez, shot Whitman twice in the head with his shotgun. As Whitman fell, Martinez grabbed McCoy's gun and continued to fire into the sniper's writhing body.

Charles Whitman was dead. It was 1.24 p.m. He had occupied the observation deck for just 96 minutes.

Police later discovered that Whitman had also killed his wife and his mother to add to the fourteen dead and dozens of injured in the shooting spree. They found what amounted to a suicide note with a bizarre request that a post-mortem be carried out on his body. An autopsy revealed a small tumour in his brain.

Charles Whitman's murders are so appalling as to be almost beyond comprehension. The killings rocked America. Not only did they bring back echoes of the Kennedy assassination in the same state, the country was still in shock from the mass murder of a group of young women at a nurses' home in Chicago just a few weeks earlier. From a sniper's point of view Whitman's accuracy is extremely impressive. It's believed he hit a target with every shot – not a bullet wasted. He concealed himself when appropriate and, even when the police started shooting from below, he never offered them a target.

Like Lee Harvey Oswald, Whitman was a Marine, and although he was given an honourable discharge and buried in a flag-draped coffin he does not appear to have been a very good one. Even so, he had been through Marine basic training and was effectively conditioned to kill. The ethos behind basic training is to break a recruit down in the first two weeks through verbal and physical abuse, exhaustion and humiliation, then remake him in the Marines' own image during the next eight weeks. It is, as one recruit put it, like living in a 'perpetual state of shock and fear'.

Those who want to think best of Charles Whitman, his friends and relatives, have grasped at the presence of the tumour as an explanation for his murderous behaviour. Medical experts believe that unlikely, an opinion borne out by the fact that our cancer wards are mercifully free of homicidal patients. What is not in doubt is that Whitman, who had been around guns from the age of two, had developed the mentality that allowed him to kill without compunction.

It is the function of the sniper to bring confusion to the enemy on the battlefield. In civilian life Charles Whitman brought terror and panic to the campus of the University of Texas for just over 90 minutes in August 1966. More than a quarter of a century later a sniper would grip the nation's capital in fear and alarm for more than three weeks.

Montgomery County is a small but wealthy suburb some fifteen miles north of Washington, DC. Not entirely crime-free – the local police department has more than a thousand men – but not a seething cauldron of criminality either. The murder rate in Montgomery County runs between 15 and 20 deaths per year. In one day in October 2001, the murder rate jumped by 25 per cent.

On 2 October, 55-year-old James Martin was shot in the back and killed as he was crossing a super-market car park. A short time earlier a bullet had passed through the large plate-glass window of a crafts store about two miles away in Aspen, Maryland. The incidents were alternately tragic and curious but of themselves unremarkable.

The following morning, as police began to investi-gate the separate incidents, wondering privately if there might be a connection, all hell broke loose. James Buchanan was mowing a neighbour's lawn on a ride-on mower when he was shot in the back and killed. Half an hour later, taxi-driver Premkumar Walekar was fatally shot in the chest at a filling station. Less than half an hour after that, at 8.37, Sarah Ramos was shot through the head as she sat on a park bench. She died instantly. A little more than two hours after James Buchanan had been shot, 25-year-old nanny Lori-Ann Lewis Rivera was fatally shot in the back as she was cleaning out her people carrier.

The Montgomery County police force was in a state of shock. Captain Nancy Demme described the killer as being like 'a ghost moving through the area' leaving a trail of death in his wake.

That night he struck again, this time in the capital itself. A 72-year-old man, Pascal Charlot, was shot as he was crossing a busy Washington street. Barely fourteen hours after starting his day's work the killer was sufficiently emboldened to have moved into the capital.

At this stage all the police could assume was that the killer was male – for no other reason than that they thought they were dealing with a serial killer, a rare avocation for a woman. They didn't know if all the shots had come from the same gun. The only clues they had were fragments of bullets taken from the shop that could be compared to fragments taken from the bodies of the victims.

By the following morning, when the Bureau of Alcohol, Tobacco and Firearms confirmed that the shots had all come from the same gun, the shooter had struck again, at another craft store in Fredericksburg, Virginia. A 43-year-old woman was shot in the back while she was loading shopping into her car. She survived. The Washington Sniper, as he was quickly tagged by the media, had missed for the first time since beginning his killing spree.

Although it made for lurid headlines, the media nickname was erroneous. As would become apparent, there was nothing in the behaviour of the shooter to suggest that he was a trained sniper; he was merely a marksman shooting from cover. However it is a significant indicator of the psychological impact of the sniper that this mystery man was immediately labelled as one of the elite breed.

After three bloody days, all the police had to go on was ballistics information. It told them that they were dealing with a shooter who was using a .22 calibre weapon, or one of a similar category, to fire high-velocity bullets. These bullets could be travelling anywhere between 3,000 and 3,500 feet per second. The chances of anyone surviving a hit from one of them were slim.

The tension in Montgomery County was rising. After an anxious but incident-free weekend the decision was taken to open the schools as usual on Monday morning, 7 October. The school gates had barely opened when the killer struck again. The eighth victim was a 13-year-old boy in Bowie, Maryland, who was shot once in the stomach. The boy was in critical condition for a time but survived. Once again the so-called sniper had failed to claim the life of his target.

Theories abounded. Was it a new wave of terrorist killings? A disgruntled veteran? Someone out for ransom? The first clue came in the schoolyard at Bowie. The killer had left a tarot card – Death – with a message. It said: 'Mr Policeman, I am God'. They also found a .223 shell casing, suggesting a high-powered weapon had been used.

The presence of the tarot card and the note was also feeding conjecture as to who the shooter might be. So-called 'death cards' were often left on the bodies of Viet Cong by American soldiers. Was the killer a Vietnam veteran? Alternatively, 'I am God' suggested a deluded serial killer, with its echoes of

Jack the Ripper's letters or the rantings of the Son of Sam. After seven shootings what was becoming more apparent however was that the police were dealing with a marksman rather than a sniper. There had been only one head shot; most of the victims had been shot in the chest or back, suggesting the shooter was not schooled in selecting the triangle of opportunity. Also, the angle of elevation on most of the victims was sufficiently low to suggest this was not the work of a trained sniper who would prefer to shoot from above. Nonetheless the label stuck.

Two days later the gunman struck again. On Wednesday 9 October, Dean Meyers was shot in the chest as he was filling his car at a petrol station. On the Friday morning Ken Bridges was shot and killed at a filling station with a policeman standing only 50 yards away. Police hoped that the Bridges slaying might offer a lead after witnesses reported seeing a white van nearby. This sparked a flood of calls from the public identifying white vans, but since there were more than 100,000 of them registered in the area it was hardly a concrete opportunity.

As on the battlefield so it was in this prosperous suburban community: fear of the sniper touched something primal. Filling stations turned their closed-circuit cameras around so that they pointed to the highways and roads rather than their forecourts; workers gathering nervously outside buildings for cigarette breaks pressed themselves as close to the concrete as possible rather than stand around in their usual huddles; and the Guardian Angels, the self-styled vigilantes who had policed the New York subway, volunteered to stand guard in public places.

None of this made anyone any more secure as they waited for the next bad news.

It came on Monday 14 October. Police had warned people to avoid wooded areas and to use covered car parks. In one such car park in Falls Church, Virginia, Linda Franklin and her husband were loading packages into their car. She was shot, and in spite of intensive medical treatment died shortly afterwards.

Four days later, the hunt took an even more bizarre turn when the sniper phoned the police. Not only did he phone them, he revealed that he had phoned them twice before. Transcripts of the later message reveal his use of the word 'we', suggesting the police faced a team of at least two. Ironically, although his marksmanship did not suggest a trained sniper he did appear, like a sniper, to be part of a two-man team. This call fared no better than the previous two occasions: the operator suggested he phone the official hot line and the shooter hung up.

The following day he shot again. This time his victim was a man getting into his car as he left a Ponderosa Steak House in Ashland, Virginia. He was shot in the stomach but not fatally. Police found a three-page letter attached to a tree berating them for their incompetence and demanding a $10 million bounty. The gunman also revealed he had tried to call them and that no one had taken him seriously, and blamed the last five shootings on the failure of the police to pay attention to his calls. In one final note of black farce, he left a number at which he could be contacted at a given time. By the time the letter had been cleared by forensics and opened the deadline had passed, forcing Montgomery County Police Chief Charles Moose to make a cryptic plea for the killer to get in touch again.

By 19 October, almost three weeks since the shooting started, there was no sign of a breakthrough. It seemed the killer could come and go at will, and on Tuesday 22 October he was back where he had started, in Montgomery County. Bus driver Conrad Johnson was fatally shot as he stood in the doorway of his vehicle. Again there was another critical letter with a chilling postscript: 'P.S. your children are not safe anywhere at any time.'

The situation threatened to spiral out of control, with an increasingly maverick media and a public rightly angry over the threats to their children. Then, in the most bizarre turn of all, the case effectively solved itself.

In his call of 18 October, the shooter had boastfully told police to check a robbery-murder at a liquor store in Montgomery, Alabama. That crime had taken place on 21 September. Two women had been shot and one of them later died. Police in Alabama had found a fingerprint at the scene, which led directly to one individual and, indirectly, to a possible accomplice. The Montgomery County authorities agonised about releasing the names, ostensibly for fear that the suspects might flee but more realistically for fear of mob rule once the names were in the public domain. The media had no such qualms. They declared that the police were looking for John Allen Williams and John Lee Malvo. They also revealed that the men were in, not a white van, but a dark blue 1990 Chevrolet Caprice. Live news bulletins carried footage of police searching a house in Tacoma in Washington State, some 3,000 miles away.

A truck driver at a rest stop in Frederick, Maryland had seen the news as well. He recognised the dark blue Caprice in the car park and called it in. By the greatest good fortune neither Williams nor Malvo had been watching television, because they were asleep in the car. They were wakened by the sound of the crack FBI Hostage Rescue team smashing in the car windows accompanied by 100 armed policemen.

John Allen Williams had changed his name to John Muhammad in 2001. He was forty-one. Malvo, the son of a woman Muhammad had dated for a time, was seventeen. When the police examined the Caprice they found a Bushmaster XM-15 .223 rifle that was quickly confirmed as the murder weapon. They also found, chillingly, that the boot of the car had been modified. There was a small hole in the trunk which could conceivably have allowed the shooter to lie down inside the car and fire through the hole – like a Western Front loophole – while the driver was free to make their getaway.

Muhammad appears to have been responsible for the majority of the killings. Like Lee Harvey Oswald and Charles Whitman he was a former Marine, and had qualified as a marksman, but he was not a sniper. The teenage Malvo, it seems, shot the schoolboy, wrote the letters, and left his voice on the tape-recorded calls to the police. The motive for their crimes remains unclear. Muhammad's ex-wife believes that it was all a cover for a plan to kill her, blame it on the 'sniper', and then regain custody of his children. Muhammad and Malvo are now awaiting trial in Virginia. They will be tried separately and, if convicted, face death by lethal injection.

Interestingly, former Vietnam sniper Ed Kugler was asked for his comments during Muhammad's reign of terror. He suggested that his motive might have been the simplest one of all: 'I can say from personal experience that it's about control, because that was what you felt as a sniper, a tremendous sense of power. Maybe the best word is "intoxicating".'

Perhaps John Muhammad and Lee Malvo just wanted to be noticed. Perhaps they wanted to be in charge for a while as they captured the attention first of America and then the world for twenty-two days. They made people look over their shoulders, they made people sprint for their cars, and they made parents fear for their children. Above all, they reminded us of our abject fear of the nameless terror that strikes when we are at our most vulnerable and then vanishes 'like a ghost'.

16

THE ART OF KILLING

The establishment of the Marine Scout Sniper School at Quantico in 1977 confirmed that sniping had finally been accepted as a permanent part of the battlefield.

Fewer than 2 per cent of the casualties in the war in Vietnam were caused by snipers: 13,000 out of a total enemy death toll of anywhere between 600,000 and more than 900,000. Sniper kills did not determine the outcome of the war. There were no Rambos, no superhero soldiers who could have won the war on their own. The true value of the sniper in Vietnam was in a reconnaissance and intelligence-gathering capacity. Huge amounts of information were gained from material recovered from the bodies of sniper kills.

What the snipers did in Vietnam was to legitimise their craft. They made commanding officers and tacticians aware of the value of long-range marksmanship. There were still those who advocated closing down the sniper schools and returning the specialised shooters to regular units after Vietnam. The Army did shut down its sniper-training unit, the Marine Corps kept theirs.

By the beginning of the 1980s most of the major armies of the world had their own dedicated sniper units. Exchange systems were set up where snipers from one country could go and train in an allied country, encouraging a free flow of ideas, tactics and information. This proved vital in combined operations, sponsored by NATO for instance.

Since Vietnam, America has continued in its role as the world's self-appointed policeman. American soldiers have seen service in Beirut, Panama, Grenada, twice in the Gulf, Somalia, the Balkans, and Afghanistan. In each of these engagements the sniper has been one of their most valuable weapons. Fighting in the deserts of the first Gulf War or in the hill country of Afghanistan is tailor-made for the sniper. Indeed, when the Russians had their own ill-fated Afghanistan adventure, among their few successes were their snipers with their powerful and deadly accurate Dragunov rifles.

During the last quarter of the twentieth century the world saw new armies being formed and new wars fought, especially in the Middle East. Inexpensive, mobile and usually highly motivated, the sniper was the ideal combatant

The Israeli Army became one of the world's best-equipped and most battle-hardened forces. The Israelis even designed their own sniper rifle; the Galil is an impressive performer in the hands of an expert and can kill almost unfailingly at 600 to 800 yards. Chuck Kramer, an American adviser in the Middle East, recalls using a Galil to deadly effect in Beirut in 1982:

I saw this big motor scooter coming towards me about a kilometre away with two guys on it. They're both carrying SKS carbines slung over their shoulders, two shopping bags full of food and these canvas carriers for the RPGs [rocket-propelled grenades] on the front of the motor scooter. I lined them up and fired by sheer instinct. It was a classic shot. It must have gone through the driver and into the passenger.

The 1970s also saw a dramatic rise in terrorism. To a minority group or oppressed nation faced with a larger and, to them, oppressive enemy, terrorism holds an obvious appeal and sniping is one of the most easily deployed tactics.

The IRA 'did great execution' among the British soldiers and civilian population of Northern Ireland during the shooting war of the Seventies. Certain areas of Belfast, such as the notorious Divis Flats, were as much a sniper's paradise as the Western Front at Neuve-Chapelle. In an attempt to combat the Republican snipers, the British Army brought in its own. These were more commonly used in a reconnaissance role, but there were times when they did what they were trained to do.

The most contentious episode in this grisly chapter of British history remains the so-called Bloody Sunday shootings in Londonderry in 1972. A civil rights march erupted in violence when members of the Parachute Regiment opened fire on the protesters. The soldiers claimed they had been shot at and were firing in retaliation. Fourteen people died.

Forensic examination afterwards found that three

of the victims – John Young, Michael McDaid and William Nash – bore what appear to be definitive sniper wounds. They had each been shot through the eye with the bullet travelling down through the neck, into the heart, and out of the back. Their wounds were typical of a victim of a gunman who understood the concept of the triangle of opportunity that is sought by every sniper. Obviously the shots came from elevated vantage points. Retired Army surgeon Hugh Thomas is on record as believing that these three men died as a result of targeted sniper fire:

> These shots could only have come from a higher level. It would be almost impossible for those three men in the few seconds available to them to bend to exactly the same angle and face exactly the same way and be shot in exactly the same fashion. It would be extraordinary.

The weapons faced by the British Army in Northern Ireland were of high quality. The masked and hooded gunmen of the IRA, the UDA, the UVF, and various other terror organisations were at the cutting edge of an incredibly sophisticated operation. Through the proceeds of robberies, drugs and prostitution rackets and misguided private donors they were incredibly well funded organisations. In some cases they were directly sponsored and funded by states that were themselves opposed to the British. As a consequence they were able to buy sniper weapons that were more sophisticated and more powerful than anything the British Army had as standard issue. For a time the IRA weapon of choice was the .50 mm Barrett Rifle, made in the USA, a weapon that can punch a hole in light armour at a distance of more than a mile. It was particularly effective in attacks against police stations and military checkpoints. The haul of guns discovered in a terrorist arms cache would make an Army supply officer weep with frustration.

Well-funded terrorists were not confined to Northern Ireland. Terrorist groups or, depending on your political ideology, freedom fighters all over the world have been able to rely on billions of dollars in funding from oil-rich states. In the days before the break-up of the Soviet Union, Colonel Gaddafi and his state-sponsored Libyan terrorists were able to draw on Moscow's political influence. There were others who were just as happy to draw on the power and influence of the Americans, including the American-trained Mujahidin in Afghanistan.

The eventual collapse of the Soviet Union – caused partly by the economic effects of excessive competition in the arms race – meant that the anticipated push-button war of ICBMs never happened. Instead there was a proliferation of small, disorderly, limited wars. The missiles launched in these conflicts were impressive ordnance, but they were not the city-destroying weapons that Cold War planners feared. Instead, at the beginning of the new millennium the role for the infantryman on the battlefield seemed more vital than ever. And where there was room for the infantryman there was the potential for the well-aimed shot to make a difference.

In the Falkland Islands in 1982 snipers were used by the British to probe the Argentine defences and test the strength of resistance before they committed to a landing. In the Gulf War of 1991, in which US President George Bush mobilised a joint Western–Arab coalition after Saddam Hussein had invaded Kuwait, snipers again came into their own in the early days of the fighting. While Bush drew his line in the sand and issued an ultimatum, Special Forces and Navy SEAL snipers were already at work infiltrating the Iraqi defences.

In the *New York Times* of 3 February 1991 reporter Chris Hedges described a group of American snipers as they prepared for the invasion of Kuwait:

The grand strategy of battle meant little to Staff Sergeant Douglas A Luebke, the non-commissioned chief of the platoon and his men – other than that for the first time they will be called upon to go into combat, to do the job many have trained for years to carry out. They will crawl unseen within yards of enemy lines and, firing three shots from an exposed position, take out enemy soldiers one by one.

'It is the art of killing,' Sergeant Luebke said. 'We have to be perfect.'

The snipers said they were trying to avoid thinking of their Iraqi opponents as men with families and children. Several said the reputed maltreatment of American prisoners of war had steeled them to their task. 'I try not to think about the other man's personal life,' Sergeant Anderson said. 'I concentrate on him being the enemy. If I were to give him sympathy I don't think I would be effective.'

The sniper would argue that each life he takes saves the life of one or more of his comrades in arms. The value of the tactical shield provided by the sniper is that the further in front of his own lines it is spread the more protection he provides for his own men. There are times when snipers go above and beyond what is required of them.

The chaotic military action that took place in the Somalian capital Mogadishu on 3 October 1993 has been well chronicled by Mark Bowden in the exemplary *Black Hawk Down*. One isolated incident stands out in a day of heroism and human tragedy. What was supposed to be a simple extraction by US Rangers of prominent Somali warlords had gone catastrophically wrong. No battle plan ever survives its first encounter with the enemy, and this was a truly hideous example of that military truism. One small mishap – a soldier injured as he rappelled out of a helicopter – begat another and another. Poor communications and lack of a clear command structure produced a domino effect. The result was a firefight lasting for fifteen hours; in battle a fifteen-minute firefight would be considered exceptional.

US Rangers were trapped in the city by a huge force of Somalis intent on revenge against the invaders of their country. Their focus was one of the two Black Hawk helicopters which had been downed during the afternoon. There were American crew members still alive in the wreckage, and without intervention they were certain to be killed.

Delta Force snipers Gary Gordon and Randy Shughart were providing aerial support from a helicopter circling overhead. As the Somalis began to swarm around the survivors sheltering in the wreckage the Delta Force men kept them at bay with an impressive display of precision shooting from a helicopter that swayed and bobbed as it tried to avoid being hit. They realised that there was only so much they could do from the air before their own helicopter was hit. Without hesitation, and knowing they were facing certain death, the two snipers volunteered to be inserted into the heart of the firefight. They rappelled down their ropes and made their way to the wreckage. Armed with pistols and rifles and whatever ammunition they could gather they put up a heroic resistance. Both men were wounded but continued to fight on until they were fatally shot. They sacrificed themselves for pilot Gary Durant, who was taken prisoner by the Somalis but ultimately released.

Master Sergeant Gary I. Gordon of Lincoln, Maine and Sergeant First Class Randall D. Shughart of Newville, Pennsylvania were both posthumously awarded the Medal of Honor, the United States highest decoration for valour.

17

THE LIVING WEAPON

Writing almost 200 years ago the Prussian military philosopher Carl von Clausewitz observed: 'Weapons affect tactics, which, in turn, determine their design.' Clausewitz's observations were based on the armies of the French Revolution and the Napoleonic Wars, at the very advent of the sharpshooter as a tactical weapon. But his comments can be just as easily applied to the sniper, in the way he has transformed the battlefield and the craft of warfare and been transformed by them in return.

The sniper has created what his commanding officers might consider to be a virtuous circle. His presence has changed military tactics, and now those tactics revolve to a great extent around him.

The combination of advances in rifle technology and the arrival of the sniper on the battlefield meant that war could not be waged as combat between two static forces. Once the sniper was mobile and deadly, tactics changed to accommodate that, reaching a peak in the American prosecution of the ground war in Vietnam. The sniper also made war personal and terrifying for the enemy. It had always been a bloody business, but up to the eighteenth century it had been possible to find some respite, a safe haven on the battlefield. The advent of the sniper meant there was no safety, even in your own lines, as thousands of British soldiers discovered on the Western Front. He had become a fearsome psychological weapon.

Once you are targeted by a sniper your chances of survival are slim. He is trained not to miss and modern technology makes it unlikely that he will. That same modern technology means he can be more than a mile away when he takes his shot. If a soldier is killed by a sniper his comrades will have no idea where the shot came from, or where the next one is going to come from. The result is panic, confusion, and blind terror.

It is no coincidence that the best snipers have almost always been recruited from the hunting fraternity. This ability to stalk and kill is what sets the sniper apart from others who are merely good shots. Police marksmen are often loosely referred to as snipers, but they are not. Snipers have a specific task. Effectively they have become weapons to be deployed like any other piece of tactical ordnance. Like every other weapon in the arsenal of the British Army, the sniper has a manual and an official description laid down in that manual:

The sniper is a selected soldier who is a trained marksman and observer, who can locate and report on an enemy, no matter how well concealed, who can stalk or lie in wait unseen and kill with one shot.

The American field manual stresses similar qualities of stealth and observation plus an additional quality of finesse:

A well-trained sniper, combined with the inherent accuracy of his rifle and ammunition, is a versatile supporting arm available to an infantry commander. The importance of the sniper cannot be measured simply by the number of casualties he inflicts on the enemy. Realisation of the sniper's presence instils fear in enemy troop elements and influences their decisions and actions . . . The sniper's role is unique in that it is the sole means by which a unit can engage point targets at distances beyond the effective range of the M16 rifle. This role becomes more significant when the target is entrenched or positioned among civilians, or during riot control missions. The fire of automatic weapons in such operations can result in the wounding or killing of non-combatants.

In an increasingly automated battlefield, the sniper is one of the few remaining warriors left in the conflict. In both Gulf Wars and in the Balkans, the thrust of the NATO and Anglo-American attacks was based on their devastating air power. Even in the tactics of 'shock and awe', the sniper has a vital part to play. It was the snipers of the SAS or SBS, or Delta Force or Navy SEALs, whose job it was to range far ahead of their own forces, alone behind enemy lines. There they could gather intelligence and lay the groundwork for bombing or missile raids by supplying precise coordinates for key targets. The twenty-first-century sniper has become the epitome of the ideal envisioned by Hesketh-Pritchard almost a century ago. With the emphasis on carefully targeted air attacks, ground forces are less at risk and casualties among them are fewer.

The sniper is the smartest of 'smart' weapons. Through the fog of war he is able to distinguish friend from foe, target from survivor, in confused surroundings. It is no coincidence that the sniper scout's role has evolved in modern warfare to include being a spotter for smart bombs. Once again the sniper's ability saves lives, a virtue stressed in the US Army definition especially when dealing with enemies taking cover among civilians. Second World War sniper Harry Furness insists this is a quality that was frequently overlooked by those who are quick to condemn the sniper as a cold and ruthless killer:

Walking around today are very many former officers and NCOs who would have been targeted and killed by enemy snipers if it had not been for what is known as 'the sniper shield'. This very efficient tactic is in constant operation in war. Dedicated sniper teams operate far beyond a regiment's foremost positions and stalk, counter, and eliminate enemy snipers. Those enemy snipers would otherwise cause devastating losses among our best battlefield leaders.

The modern sniper is perhaps the most versatile soldier on the battlefield. He has emerged from the press of the throng of infantrymen as a warrior who stands alone. He is totally self-reliant, living off the land and his wits, the most vital link in the chain of combat. Ranging at will over the battlefield, he will push himself to his technological and psychological limit to provide protection for his comrades, intelligence for his commanders, and deadly interdiction whenever it is necessary. He is essentially a living weapon. Fast. Mobile. Independent.

On 13 June 2003 American forces launched Operation Desert Scorpion across central and northern Iraq. The aim of this major initiative was to root out remaining pro-Saddam loyalists who were believed to have been organising an anti-American resistance in the wake of the Second Gulf War. Desert Scorpion was only hours old when it suffered its first fatalities, among them Private Shawn Pahnke, of the 1st Brigade, 1st Armored Division. The 25-year-old was sitting in the back of a military Humvee on patrol in Baghdad when he was shot in the back by a sniper. He had been wearing a flak jacket but the shooter had chosen an angled shot that would miss the armour plate in the jacket. The bullet tore through the padding, drilled into his back, and Pahnke was dead by the time medical aid arrived. The sniper was not found, nor is he likely to be.

Unlike the death of Lord Brooke at Lichfield, the shooting of Shawn Pahnke did not change the course or the outcome of any battle. But more than 400 years later it did prove the invincibility of the sniper. Even the American Army – the most technologically superior military force assembled in history – has no protection against the most basic and deadly of strategies: one shot, one kill.

BIBLIOGRAPHY

Peter Abbot, Nigel Thomas and Mike Chappell, *Germany's Eastern Front Allies 1941–45* (Osprey 1998)

Stephen E. Ambrose, *Citizen Soldier* (Simon & Schuster 1997)

—, *D-Day* (Touchstone 1994)

Albert Axell, *Russia's Heroes 1941–45* (Constable 2001)

Charles Baldwin, *African Hunting* (Books of Bulawayo 1981)

Michael Barthrop and Pierre Turner, *The British Army on Campaign*, vols 1 & 2 (Osprey 1996 & 2001)

Antony Beevor, *Stalingrad* (Viking 1998)

Joanna Bourke, *An Intimate History of Killing* (Granta 1999)

Mark Bowden, *Blackhawk Down* (Corgi 2000)

Peter Brookesmith, *Sniper* (St Martin's Press 2001)

Bob Carruthers, *The English Civil Wars* (2000)

Cassell's Biographical Dictionary

Bruce Caton and James M. McPherson, *The American Heritage New History of the Civil War* (Viking 1996)

Concise Dictionary of American History, The (Charles Scribner's 1983)

John J. Culbertson, *A Sniper in the Arizona* (Ivy Books 1999)

Kenneth C. Davis, *Don't Know Much about History* (Avon Books 1995)

William C. Davis, *The Orphan Brigade: The Kentucky Confederates Who Couldn't Go Home* (Doubleday 1980)

Captain J.C. Dunn, *The War the Infantry Knew* (Abacus 2001)

Colonel Trevor N. Dupuy, *The Evolution of Weapons and Warfare* (Da Capo 1990)

Ron Field and Robin Smith, *Uniforms of the American Civil War* (Brassey's 2001)

Douglas Southall Freeman, *Washington*, abridged in 1 vol. by Richard Harwell (Macmillan 1968)

J.F.C. Fuller, *Armament and History* (Da Capo Press 1998)

General John R. Galvin, *The Minute Men* (Pergamon-Brassey 1984)

David M. Glantz and Jonathan M. House, *When Titans Clashed* (Birlinn 2000)

Adrian Goldsworthy, *Cannae* (Cassell 2001)

J. Glenn Gray, *The Warriors: Reflections of Men in Battle* (Bison Books 1998)

Paddy Griffith, *Rally Once Again: Battle Tactics of the American Civil War* (Marlborough 1989)

Philip J. Haythornthwaite, *The World War One Sourcebook* (Arms & Armour 1992)

Charles Henderson, *Marine Sniper* (Berkeley 1986)

H. Hesketh-Pritchard, *Sniping in France* (Hutchinson 1920)

Richard Holmes, *Redcoat* (HarperCollins 2002)

Robert Underwood Johnson, *Battles and Leaders of the English Civil War* (Castle, 1990)

Thomas Keneally, *Confederates* (Sceptre 1988)

Ed Kugler, *Dead Center* (Ivy Books 1999)

Michael Lee Lanning, *Inside the Crosshairs: Snipers in Vietnam* (Ivy Books 1998)

Jon E. Lewis (ed.), *The Mammoth Book of Soldiers at War* (Robinson 2001)

Lynn Macdonald, *1915: The Death of Innocence* (Headline 1994)

Heinz K. Meier (ed.), *Memoirs of a Swiss Officer in the American Civil War* (Herbert Lang 1972)

Oxford Companion to Military History, The, ed. Richard Holmes (OUP 2001)

Martin Pegler and Ramiro Bujeiro, *The Military Sniper since 1914* (Osprey 2001)

Stephen Pope and Elizabeth-Anne Wheal, *The Macmillan Dictionary of the First World War* (Macmillan 1995)

Ernie Pyle, *Brave Men* (Bison Books 2001)

Recollections of Rifleman Harris, The, ed. Christopher Hibbert (Windrush Press 1996)

Laurence Rees, *Horror in the East* (BBC 2001)

Stuart Reid and Richard Hook, *British Redcoat 1740–1793* (Osprey 2001)

Guy Sajer, *The Forgotten Soldier* (Cassell 1999)

Peter R. Senich, *U.S. Marine Corps Scout Sniper* (Paladin Press 1993)

—, *Limited War Sniping* (Paladin Press 1977)

Stuart Sifakis, *Who Was Who in the Civil War* (Facts on File 1988)

Ian Skennerton, *The British Sniper – British and Commonwealth Sniping and Equipment 1915–1983* (Skennerton 1983)

Sniping, Scouting and Patrolling, A Textbook for Instructors and Students (Gale & Polden)

Mark Spicer, *Sniper: the Techniques and Equipment of the Deadly Marksman* (Salamander Books 2001)

James L. Stokesbury, *A Short History of the American Revolution* (Quill 1991)

William Taylor, *Student and Sniper-Sergeant: A Memoir of J.K. Forbes* (Hodder & Stoughton 1916)

Times History of War, The (Times Books 2000)

Geoffrey C. Ward, Ken Burns and Ric Burns, *The Civil War* (Knopf 1990)

Joseph T. Ward, *Dear Mom: A Sniper's Vietnam* (Ivy Books 1991)

C.V. Wedgwood, *The King's War 1641–47*

F. Cowley Whitehouse, *The Sniper: a Book for Boys* (James Nisbet 1907)

Mike Wright, *What They Didn't Teach You about the American Revolution* (Presidio Press 2001)

Peter Young, *The Great Civil War 1642–1648* (Spurbooks 1980)

—, *Peter Atkyns* (Longmans 1967)

Steven J. Zaloga and Ronald B. Volstad, *The Red Army of the Great Patriotic War* (Osprey 1997)

ACKNOWLEDGEMENTS

I have been fortunate in my writing career in that one book has led more or less to another. This book has its origins in the research for my last book, *Dynamo*, during which I came across the story of Vasily Zaitsev and the Soviet snipers.

I have also been fortunate in my writing career in finding people sufficiently kind and generous with their time and advice to steer me in the right direction. I would like to thank Professor Richard E. Holmes for early encouragement as well as Ministry of Defence staff and members of the armed forces – past and present – for their help. My thanks also to Hugh Thomas for providing me with a brief but fascinating introduction to wound ballistics.

This book could not have been completed without the help of librarians and archivists all over the world. I am especially grateful to staff at the Imperial War Museum in London, the online staff at the Library of Congress, and the Mitchell Library in Glasgow. I would also like to thank Liz Howell for her material on the First World War.

My thanks as always to my agent Jane Judd for not

allowing me to be deflected from the task at hand and to Clive Priddle and Nick Davies at Fourth Estate for an incredibly helpful edit.